Perspectives in Mössbauer Spectroscopy

Perspectives in Mössbauer Spectroscopy

Proceedings of the International Conference on Applications of the Mössbauer Effect, held at Ayeleth Hashahar, Israel, August 28-31, 1972

S. G. Cohen
The Hebrew University
Jerusalem

and

M. Pasternak
Tel Aviv University and
Soreq Nuclear Research Centre
Yavne, Israel

PLENUM PRESS • NEW YORK-LONDON • 1973

Library of Congress Catalog Card Number 72-97398

ISBN-13: 978-1-4615-8689-0 e-ISBN-13: 978-1-4615-8687-6
DOI: 10.1007/ 978-1-4615-8687-6

Softcover reprint of the hardcover 1st edition 1973
A Division of Plenum Publishing Corporation
227 West 17th Street, New York, N.Y. 10011

United Kingdom edition published by Plenum Press, London
A Division of Plenum Publishing Company, Ltd.
Davis House (4th Floor), 8 Scrubs Lane, Harlesden, NW10 6SE,
London, England

PREFACE

This book is for the most part made up of the invited papers presented at the International Conference on Applications of the Mössbauer effect, held at Ayeleth Hashahar, Israel, in August 1972.

The purpose of the conference was to review perspectives in Mössbauer spectroscopy from the point of view of applications in various fields of science - mainly physics, chemistry, and biology. It was hoped that the bringing together of scientists who work in different disciplines but use a common tool would encourage interdisciplinary research and widen the horizon of the individual researcher. Both these features have been an important and unique characteristic of Mössbauer research in the past decade. Accordingly, the main papers were presented in the following areas: magnetic properties of solids, applications to inorganic chemistry, biochemical studies, amorphous systems, metal alloys, advances in high-resolution studies, and the new techniques for studying recoil-free spectra from isolated atoms in inert materials. A special panel discussion was devoted to the subject of phase transitions. Recent years have witnessed a rapid growth in the use of X-ray photoelectron spectroscopy for the study of the physics and chemistry of solids. A report on this area of research and its relation to Mössbauer studies has been included.

Over 90 contributed papers were submitted to the conference. Time permitted the actual presentation of only very few. However, a list of all the titles of the contributed papers is included in this book. This list, we believe, serves to demonstrate the ever-widening interest in the use of Mössbauer spectroscopy.

The editors have intentionally retained the informal (and sometimes even racy) style of Hans Frauenfelder in his concluding lecture, Mössbauer Effect - "Quo Vadis?" This was a summing-up of the conference, but also contains much wisdom of general validity. We hope that this contribution will also convey something of the atmosphere at the meeting, which we believe was positively affected by the "spirit of the place" - Ayeleth Hashahar. This place, Ayeleth Hashahar ("morning star") has a very poetic name, and is also near a famous archaeological site, Hatzor, which we believe was the inspiration for Michener's book "The Source."

We wish to take this opportunity to thank the following sponsoring organizations, without whose help and encouragement the organization of the conference would have been impossible: International Union of Pure and Applied Physics; The European Physical Society; The Israel Academy of Sciences and Humanities; Soreq Nuclear Research Center, Yavne; The Hebrew University of Jerusalem; Negev Nuclear Research Center - Beer Sheva; Technion, Israel Institute of Technology, Haifa.

We would also like to thank the members of the organizing committee, U. Atzmoni, J. Danon, H. Frauenfelder, U. Gonser, R. H. Herber, G. M. Kalvius, S. L. Ruby, H. Schechter, and H. DeWaard.

S. G. Cohen
The Hebrew University
Jerusalem

M. Pasternak
Tel-Aviv University and
Soreq Nuclear Research Center

November 1972

CONTENTS

INVESTIGATION OF MAGNETIC MOMENTS IN METALS BY THE MÖSSBAUER EFFECT

S. Hüfner and P. Steiner

IV. Physikalisches Institut, Freie Universität

1000 Berlin 33, Boltzmannstr. 20, Germany

ABSTRACT

The state of the art of our understanding of magnetic moments will be reviewed with special reference to contributions made by the Mössbauer effect in that field. In addition some recent results on Mössbauer measurements of Fe:$\underline{Ni_3Ga}$ and Fe:$\underline{Ni}$ will be presented.

The problem of the formation and behavior of magnetic moments in metals may be sketched as follows /1/. If one wants to know what the magnetic moment of a free trivalent iron ion is, one can do this simply by applying Hund's rule. This shows that the ground state of Fe^{3+} is $^6S_{5/2}$ and the magnetic moment of that state is $5\mu_B$. The problem is equally simple for Fe^{3+} in a diamagnetic crystal like Al_2O_3. In that case the atomic energy levels are modified by the crystal field, but still, the ground state is the same as in the free ion and the magnetic moment therefore is $5\mu_B$ as has been confirmed e.g. by ESR measurements. The next step in that sequence is to determine the magnetic properties of a crystal like Fe_2O_3. Here one has of course magnetic interactions between the iron ions which make things more complicated. But still a high temperature susceptibility measurement shows that the magnetic moment is $5\mu_B$. The problem gets really complicated if one wants to know the magnetic moment of iron in iron metal. Its magnitude is $2.2\mu_B$, distinctly different from the above mentioned cases. If one persues the same way as in the case of insulators and investigates the problem of diluted iron impurities in copper, new problems arise, which are connected with the Kondo effect /1/, and that is where the problem stands at present.

In a discussion of actual experiments we shall start out with rare earth metals. This is mainly the case because for these elements the magnetic electrons are much more localized than for the transition metal elements and therefore the understanding seems simpler.

The discussion will be started with a very simple case, namely erbium metal /2/, which is a ferromagnet at low temperatures. A spectrum of erbium metal for a rotational transition with a single line source shows a 5 line-pattern, typical for a combined magnetic and quadrupole interaction of an I=2 state (Fig. 1). Assuming that one knows the nuclear moments, one can deduce from that spectrum the magnitude of the hyperfine field and the field gradient.

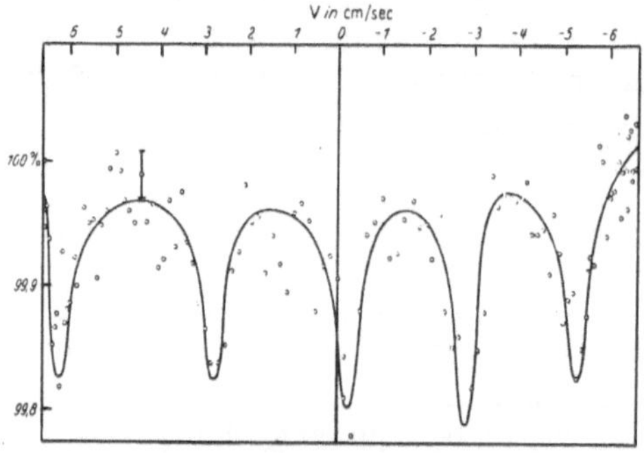

Fig. 1: Mössbauer absorption spectrum of an erbium metal absorber with a single line source ($HoAl_2$).

For an interpretation of these measured quantities, one may start out with the simplest approximation, namely that the erbium ion is not influenced at all by the crystal field and the conduction electrons and that the exchange interaction can be represented by a magnetic field. This predicts that the lowest state of the 4f configuration has a total angular momentum of $M_J = -15/2$ and the hyperfine field and quadrupole interaction of that state can be taken as equal to those measured for a free atom. This yields a magnetic hyperfine field of 7.6 Moe in very good agreement with the actually measured value. The analysis of the quadrupole interaction is more involved because it needs the knowledge of several Sternheimer shielding and antishielding factors. But taking here those derived from calculations, one is indeed able to deduce a very reasonable estimate for the actual crystal field strength. This analysis yields a total crystal field splitting of the ground state of approximately 50 ^{o}K. All these results together give strong evidence that the magnetic moment in a rare earth metal is indeed very well localized and similar in magnitude to that expected for a free ion.

Only recent neutron diffraction measurements have shown that indeed the radial distribution of the 4f electrons is different in an insulator and in a metallic compound but only by a small amount /3/.

To get more detailed insight into the conduction electron influence on the hyperfine parameters in rare earth metals, one has obviously to look into cases, where there is only a very small orbital contribution to the electron wave function, namely Eu and Gd /4/. Gd at a first sight looks again very discouraging because the hyperfine fields in Gd metal and the free Gd ion or Gd in an insulating host are very similar, namely -340 kOe. So one is left with Eu, where there is about 100 kOe difference in hyperfine field between Eu in an insulating host (-340 kOe) and in Eu metal (-265 kOe).

The hyperfine field in a metal has two additional contributions, if one compares it with that in an insulating host. These two additional contributions come from the polarization of the conduction electrons by the own 4f moment and by the 4f moments of the neighbors. One can now study the relative importance of these two contributions by diluting Eu (or Gd) with ions that have the same electronic properties as Eu (or Gd) but carry no magnetic moment. If one dilutes Eu metal with ytterbium, the hyperfine field at the Eu nucleus is changed by the amount that is produced by the polarization of the conduction electrons through the neighboring magnetic moments. It is seen that the hyperfine field increases approximately linearly from -265 kOe to -150 kOe if one goes from pure Eu metal to diluted Eu in ytterbium metal. This of course means that the neighbors produce a negative hyperfine field at the site of the Eu ion. Assuming that the core polarization contribution to the hyperfine field in Eu metal is the same as e.g. for Eu^{3+} in CaF_2, we find that the conduction electron contribution at the own site of the moment is about +200 kOe and that the integral contribution of all the neighbors is about -100 kOe. Especially the large positive contribution produced by the own magnetic moment has interesting consequences. This implies a predominantly positive, which means ferromagnetic, exchange interaction between the magnetic moment and the conduction electrons, which in turn means that the conduction electrons are predominantly of s-character. This is in very good agreement with the recent findings of ESR measurements of diluted rare earth impurities in various noble metal hosts /5/, which all yield a positive g shift and therefore a positive exchange interaction. The negative exchange interaction for rare earth moments in metallic environments as deduced from earlier ESR measurements seems therefore to be due to an incorrect interpretation of these measurements.

The same mechanism is obviously also responsible for the apparent lack of conduction electron contribution to the hyperfine field in Gd metal. Here the polarization contribution from the own moment and from the neighboring moments are roughly equal, so that the hyperfine field in the metal (-325 kOe) is almost equal to that of Gd in a diamagnetic host (-340 kOe).

The situation is somewhat more complicated in iron although there are much more experiments available in that case. First of all there has been the question whether iron is an itinerant ferromagnet or a magnet with localized magnetic moments. We think the fact that one measures a magnetic hyperfine field and that the neutron form factor in iron metal is very similar to that calculated for a free iron atom are good evidence for considerable local character of the magnetic moment in iron metal. The extensive Mössbauer effect measurements of Stearns /6/ and also of other groups have given a detailed, but still disputed picture of the conduction electron polarization in iron metal (Fig. 2). For these measurements non-magnetic atoms like silicon or aluminum are introduced into iron metal and the change of hyperfine field produced by replacing a particular neighbor of an iron atom is then observed. Due to the high resolution of the iron spectra, the effects of the neighbor shells up to the 9th nearest neighbor have been observed. The Mössbauer effect of iron diluted with aluminum and silicon shows that the contribution of the first and second nearest neighbors is negative and that of the third nearest neighbor positive. Unfortunately, these results have not

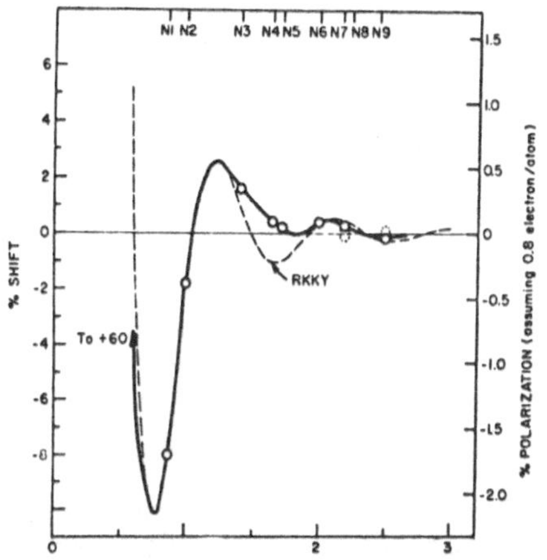

Fig. 2: Percentage shift of hyperfine fields in Fe due to diamagnetic neighbors as deduced from a careful study of the FeSi system (Stearns, Ref. 6).

been confirmed yet by NMR measurements on the same samples /7/. The NMR measurements indeed give a negative first and third neighbor contribution and a positive second neighbor contribution, which would indicate a spin density oscillation with a much shorter period.

To understand more fully the behavior of a 3d moment in a metal, one might think that it is a good idea to investigate its properties as a diluted impurity in a nonmagnetic metal. The classical system here is iron in copper which has been investigated intensively. Of course it is well known by now that here additional problems which have not yet been resolved arise, which carry the name Kondo effect /1/. Macroscopically the Kondo effect shows up as a well known minimum in the resistivity. Microscopically it can be investigated by Mössbauer effect, NMR, and neutron diffraction, but these experiments do not yet give a full picture of the phenomena. Looking at the Kondo effect in a very naive way, it arises at low temperatures by the increasing amplitude of the conduction electron scattering by the magnetic moment. This also reduces the magnitude of that moment which can easily be seen by Mössbauer effect measurements. We briefly describe here the classical experiment of Frankel et al. /8/. They have measured the dependence of the hyperfine field of iron in copper for various external fields as a function of temperature (as shown in Fig. 3). If the iron moment would behave like a classical free spin, the hyperfine field would only be a function of the ratio of the external field over the temperature and the behavior could be described by a simple Brillouin function. This is not the case. The saturation value of the hyperfine field is a function of the externally applied field. This shows the destruction of the Kondo state by the external field. If one plots the saturation field as a function of the applied field, it looks as if this indi-

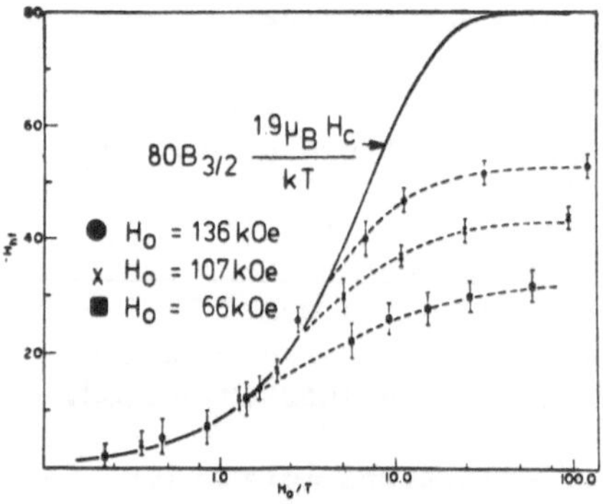

Fig. 3: Hyperfine field in Fe:Cu as a function of H/T for different values of external field H. The saturation field decreases with decreasing external field (Ref. 8).

cates a zero saturation field for zero external field. This would mean that the ground state of the Kondo system is a true singlet system.

Intensive interpretation of the various macroscopic and microscopic measurements by Heeger /1/ has supposed that a polarization cloud exists around the local moment with opposite polarization, thus producing the zero moment. Yet recent neutron diffraction measurements /9/ have not been able to confirm this interpretation.

This brief summary has shown what kind of information Mössbauer effect measurements have given to our understanding of magnetism in metals. In what follows we shall briefly deal with some aspects of local moments which we have investigated in our own laboratory recently.

The system of iron in Ni_3Ga is the one with the strongest exchange enhancement known so far /10/. Susceptibility measurements have shown that for less than 0.1% iron in this host the magnetic moment is around $40\mu_B$ and the Curie temperature is still between 30 ^{o}K and 40 ^{o}K. Extensive Mössbauer effect measurements in an external magnetic field have been made by Maletta /11/. He used a source of Co^{57} diffused into Ni_3Ga with an iron content of 20 ppm and 250 ppm. He measured the hyperfine field as a function of external field and temperature and could fit all his data with a Brillouin function. Assuming a g value of 2, he deduced the magnetic moment of the iron atoms which turned out to be $60\mu_B$ for the most dilute sample. In addition he determined the saturation hyperfine field for iron as 223 kOe.

We have made measurements with absorbers of iron in Ni_3Ga and want to discuss these results now. Our spectra clearly show two hyperfine fields and we associate them with two different crystallographic positions for the iron atoms and name them as A and B sites. The concentration dependence of the hyperfine field is shown in Fig. 4. The hyperfine field at the A site extrapolated to zero concentration is 223 kOe and corresponds exactly to that found by Maletta /11/. He did not find a field for the B site which may be due to the fact that he is working with more dilute samples.

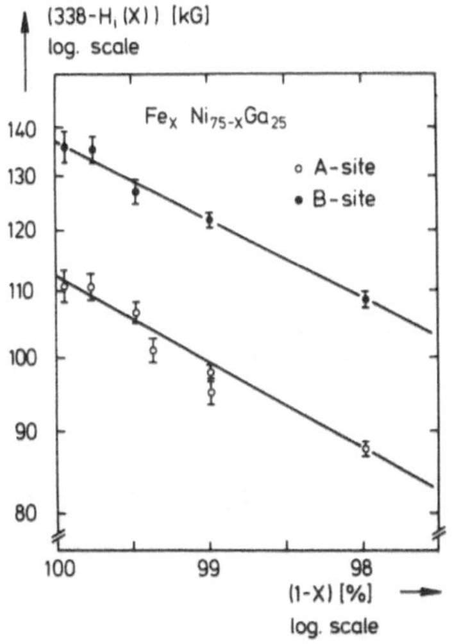

Fig. 4: Double logarithmic plot of the hyperfine field in $Fe_xNi_{75-x}Ga_{25}$ as a function of 1-x to check the functional dependence suggested in Ref.12.

Craig et al. /12/ have proposed a simple model, which describes the concentration dependence of the hyperfine field in an exchange enhanced lattice which they applied to iron in palladium. We have tentatively applied the same model to our system and this then yield the following formula for the concentration dependence of the hyperfine field:

$$H_i = H_1 + (H_o - H_1) (1 - x)^n$$

where H_1 and H_o are the fields for x=1 and x=0. A double logarithmic plot of the concentration dependence of the hyperfine field gives for both sites a straight line and the exponents for both sites are n = 12, which corresponds to the number of nearest neighbors in this compound. The isomer shift measurements show a larger isomer shift for the ions on the A sites than on the B sites. Also the hyperfine field is larger on the A sites than on the B sites. Therefore it is reasonable to associate the A site with a gallium place, where all nearest neighbors are nickel atoms and the B sites with a nickel place where some of the neighbors are galliums. In addition, one can analyze the known magnetic moments in terms of a model which assumes a polarization distribution of Gaussian shape around each ion where the maximum of the magnetization is a quarter of that of nickel. That leads to a very reasonable distribution of the magnetization, namely 6 Å for 0.1% iron in Ni_3Ga and 9 Å for 0.05% iron in Ni_3Ga.

In the last part we will demonstrate in an example how the dynamical behavior of a local moment can be studied by Mössbauer experiments. We investigated a sample Fe:Ni (with 1 at% ^{57}Fe) in the critical region below and above the Curie temperature T_c. Near T_c linebroadening in the Mössbauer spectra due to relaxation processes was observed. The parameter describing the broadening of the nuclear transitions is proportional to the time correlation function of the electronic spins. The relaxation parameter which was obtained from a fit to the Mössbauer spectra near T_c seemed to be nearly symmetric to T_c. In a small temperature range near T_c with $0.1\ ^{o}K \leq |T-T_c| \leq 2\ ^{o}K$ the relaxation parameter could be fitted to a relation: $\Delta\Gamma = \Delta\Gamma_o |T-T_c|^n$ with $\Delta\Gamma_o = 0.46(6)$ mm/s, $n = -0.65(3)$ (Fig. 5).

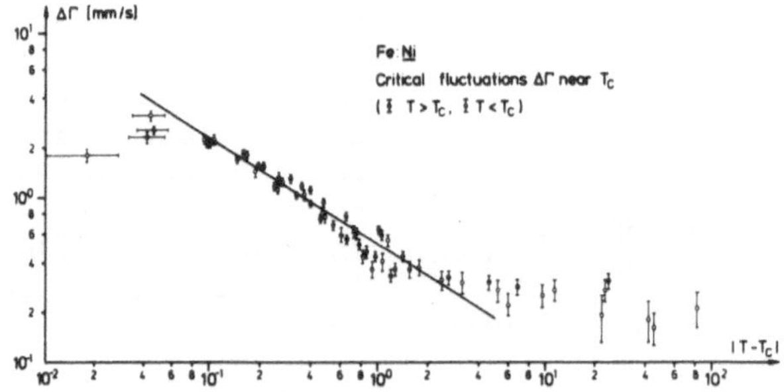

Fig. 5: Critical fluctuations in Fe:Ni near T_c determined from the linebroadening of the Mössbauer spectra.

The relaxation parameter is proportional to the correlation time τ_c. So we get: $\tau_c = \tau_o \cdot |T-T_c|^n$ with $\tau_o = 3.7(8) \cdot 10^{-10}$ sec.

The critical exponent $n = -0.65(3)$ from these data is in good agreement with a recent value $n = 0.5(1)$ of Hohenemser and Reno on the system ^{100}Rh:Ni using time differential perturbed angular correlation technique.

Due to theories of dynamical scaling /15/ this critical exponent should be related to the critical exponents γ and ν of the static susceptibility and the correlation length in the form $n = -\gamma + \nu/2$.

Using an experimental value $\gamma = 1.32(2)$ we get $\nu = 1.3(1)$. in agreement with neutron diffraction data /14/ which yield $\nu = 1.0(2)$ and considerably larger than the value $\nu = 0.74$ or $\nu = 0.65$ as predicted by a spin 1/2 Heisenberg or Ising model. This disagreement may be due to the contribution of the conduction electrons to the local moment.

Parts of this work were supported by the Bundesministerium für Bildung und Wissenschaft.

LITERATURE

/1/ For a review of the field of magnetic moments in metals see e.g. A.J. Heeger, Solid State Physics 23, Academic Press, New York (1969)

/2/ S. Hüfner, P. Kienle, W. Wiedemann, and H. Eicher, Z. Physik 182, 499 (1965)

/3/ W.C. Koehler, R.M. Moon, J.W. Cable, and H.R. Child, J. Physique 32, 296 (1971)

/4/ S. Hüfner and J.H. Wernick, Phys. Rev. 173, 448 (1968)

/5/ e.g. D. Davidov, R. Orbach, C. Rettori, D. Shaltiel, L.J Tao, and B. Ricks, Solid State Comm. 10, 451 (1972)

/6/ M.B. Stearns, Phys. Rev. 141, 439 (1966)
Phys. Rev. B4, 4069 (1971)

/7/ E.F. Mendis and L.W. Anderson, Phys. stat. sol. 41, 375 (1970)

/8/ R.B. Frankel, N.A. Blum, B.B. Schwartz, and D.J. Kim, Phys. Rev. Lett. 18, 1051 (1967)

/9/ C. Stassis and C.G. Shull, Phys. Rev. B5, 1040 (1972)

/10/ F.R. de Boer, C.J. Schinkel, and J. Biesteroos, Phys. Lett. 25A, 606 (1967)

/11/ H. Maletta, Z. Physik 250, 68 (1972)

/12/ P.P. Craig, D.E. Nagle, D.E. Steyert, and R.D. Taylor, Phys. Rev. Lett. 9, 12 (1962)

/13/ C. Hohenemser and R.C. Reno, presented of the 1971 Chicago Conference on Magnetism and Magnetic Materials

/14/ V.J. Minkiewicz, M.F. Collins, R. Nathans, and G. Shirane, Phys. Rev. 182, 624 (1969)

/15/ B.I. Halperin and P.C. Hohenberg, Phys. Rev. 177, 952 (1969)

SPIN ORIENTATION DIAGRAMS IN RARE EARTH CUBIC COMPOUNDS

U. Atzmony and M.P. Dariel

Nuclear Research Centre - Negev

and E.R. Bauminger, D. Lebenbaum,
I. Nowik and S. Ofer

The Hebrew University, Jerusalem, Israel

SPIN ORIENTATION DIAGRAMS

Whenever two isostructural magnetically ordered systems A and B form a series of continuous solid solutions A_xB_{1-x}, a spin orientation diagram may be plotted. Spin orientation diagrams are graphical representations of the directions of the preferred axes of magnetization of the A_xB_{1-x} systems in the (x, composition, T, temperature) plane. The solid solutions may exhibit a phase transition in which the easy axis of magnetization reorients itself from one symmetry axis of the crystal to another upon changing either the composition or the temperature. The (x,T) plane will then be subdivided into several regions each of which corresponds to the direction of magnetization being parallel to one of the symmetry axes of the crystal. The area above the line of the magnetic ordering temperatures corresponds to the paramagnetic behavior of the crystal.

The nature of the spin reorientation has been studied in magnetically ordered metallic solid solutions or compounds by a variety of experimental methods(1). Potentially suitable systems for spin orientation diagram determinations are the orthoferrites ($RFeO_3$), the orthochromites ($RCrO_3$) or the rare-earth garnets. The Mossbauer effect has been used to study the effect of the temperature and of the rare-rarth composition on the easy axis of magnetization in rare-earth iron garnets, though no spin orientation diagrams were actually plotted (2). The analysis of the data in these compounds has followed phenomenological lines.

In the present paper we shall describe and analvze spin orientation diagrams of ternary $R^2_x R^2_{1-x} Fe_2$ rare-earth iron Laves phase compounds. The analysis of the experimental results is based on first principles.

All RFe_2 compounds belong to the cubic Laves phase family (type $MgCu_2$) and for R = Sm through Tm have magnetic ordering temperatures close to or above 600 K. Even though all these compounds have an identical crystallographic structure, they possess different preferred axes of magnetization. This easy axis of magnetization $\vec{n}$ is parallel to the [100] direction for R = Ho,Dy, it is parallel to the [111] direction for R = Tb,Er,Tm, and Y at all temperatures up to the magnetic ordering temperature. In $SmFe_2$, $\vec{n}$ is parallel to [110] up to about 180 K where it reorients itself parallel to the [111] direction (3).

All these results have been deduced from Mossbauer effect studies on the 14.4 KeV γ ray transition of ^{57}Fe (4). Three main different types of spectra, as shown in Fig.1, were obtained for compounds having different preferred axes of magnetization. The connection between the easy axis of magnetization of the cubic Laves compounds and their Mossbauer spectra is as follows: The rare-earth ions are situated on a diamond sublattice and have a cubic site symmetry. The iron ions lie on a corner sharing network of regular tetrahedra and have a $\bar{3}m$ local symmetry, with the threefold axis parallel to the <111> directions. The electric field gradient (EFG) (excluding the small contribution due to the induced EFG) at the iron nuclei is axially symmetric and its axis is parallel to the local threefold axis of symmetry. Spectra of type a in Fig.1 are characterized by a simple six line pattern and are obtained when $\vec{n}$, the easy axis of magnetization is parallel to the [100] direction and forms an angle of $\theta = 50°44'$ with all four <111> directions. All iron ions are equivalent and give the simple six line pattern. When $\vec{n}$ is parallel to a [111] direction it forms an angle $\theta = 0$ with the axis of the EFG acting on one iron nucleus of each tetrahedron and

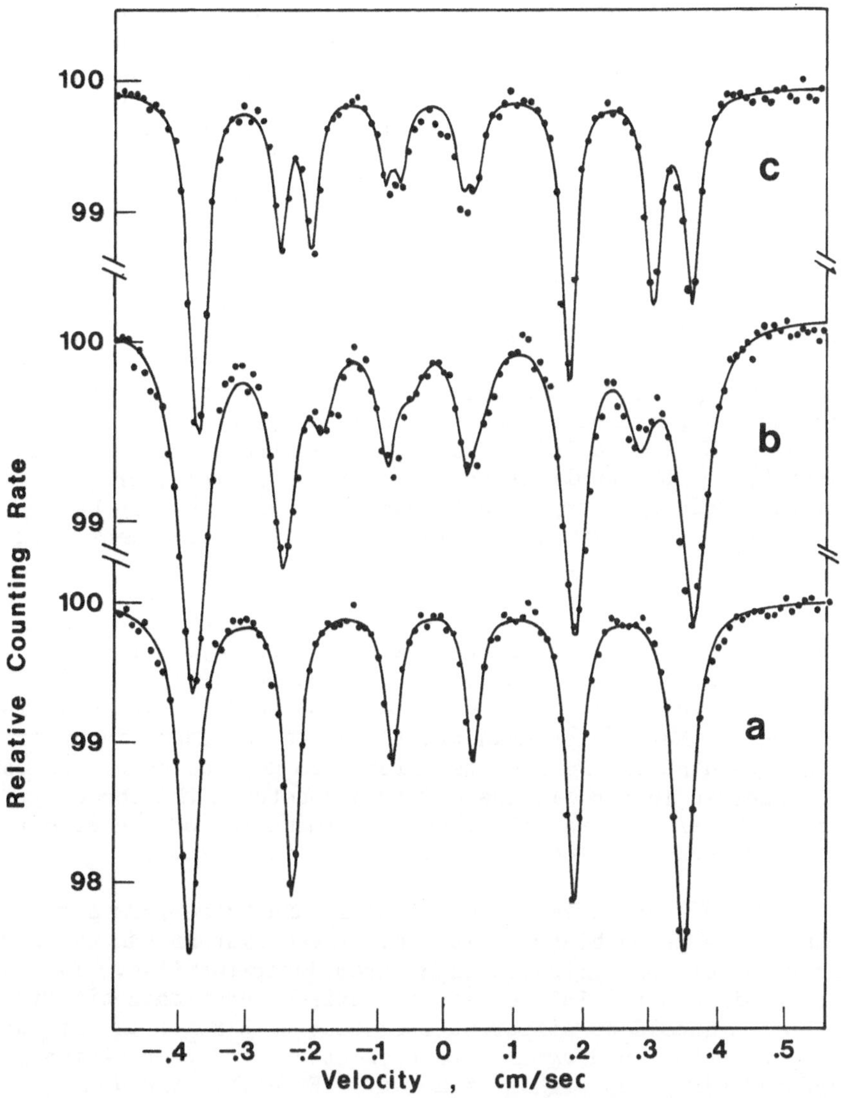

FIG. 1. Characteristic Mossbauer spectra of cubic Laves compounds having their easy axes of magnetization parallel to the major axes of symmetry. a. $Ho_{0.8}Er_{0.2}Fe_2$, $\vec{n}$ parallel to [100], simple six line pattern; b. $Ho_{0.3}Er_{0.7}Fe_2$, $\vec{n}$ parallel to [111], superposition of two six line patterns, with intensity ratio 3:1; c. $Ho_{0.75}Er_{0.25}Fe_2$, $\vec{n}$ parallel to [110], superposition of two six line patterns of equal intensity. All spectra were taken at 4.2 K.

an angle of $\theta = 70°32'$ with the axes of the EFG of the three remaining iron nuclei of the tetrahedron. There are thus two inequivalent iron sites with population ratio of 3:1 giving rise to spectra of type b, Fig. 1, which are superpositions of two six-line patterns with relative intensities in the same ratio. With $\vec{n}$ pointing in the [110] direction, it forms an angle $\theta = 35°15'$ with the axes of the EFG acting on two iron nuclei of each tetrahedron and an angle $\theta = 90°$ with the axes of the EFGs at the other two iron nuclei. Again there are two inequivalent iron sites with population ratio 2:2 giving the spectra of type c, which are superposition of two six line patterns of equal intensities.

In crystals of cubic symmetry the easy axes of magnetization lie parallel to one of the major axes of symmetry. Whenever $\vec{n}$ is parallel to any other arbitrary direction (as a result of the application of an external field, or to loss of the cubic symmetry) the 4 iron sites will usually be inequivalent and spectra different from the above mentioned 3 types will be obtained.

The hyperfine interaction at the Fe nucleus depends on the angle θ between $\vec{n}$ and the axes of local symmetry, through: 1. The contribution of the quadrupole interaction to the hyperfine splittings, which depends on θ; 2. The contribution of the dipolar magnetic fields to the magnetic hyperfine field at the Fe nucleus which also depends on θ. As can be observed in Fig. 1, the difference between the three types of Mossbauer spectra allows the determination of the direction of easy magnetization by simple visual inspection.

Two binary Laves compounds R^1Fe_2 and R^2Fe_2 form a continuous pseudobinary solution. These systems can therefore be used for spin orientation diagram determinations. In order to obtain non-trivial results the axis of easy magnetization of R^1Fe_2 should be different from that of R^2Fe_2. In this work spin orientation diagrams are presented for 5 pseudo-binary systems $R^1_xR^2_{1-x}Fe_2$ with R^1 = Ho, Dy; R^2 = Tb, Er and $Ho_xTm_{1-x}Fe_2$. About 60 samples of different compositions were prepared by arc melting followed by a heat treatment at elevated temperature. All samples were checked by X-ray powder diffraction. The lattice parameters were assumed to obey a linear relationship (Vegard's law) in the pseudo-binary solid solution. Good agreement between nominal composition and that deduced from the lattice constant measurements was observed. About one half of the samples were

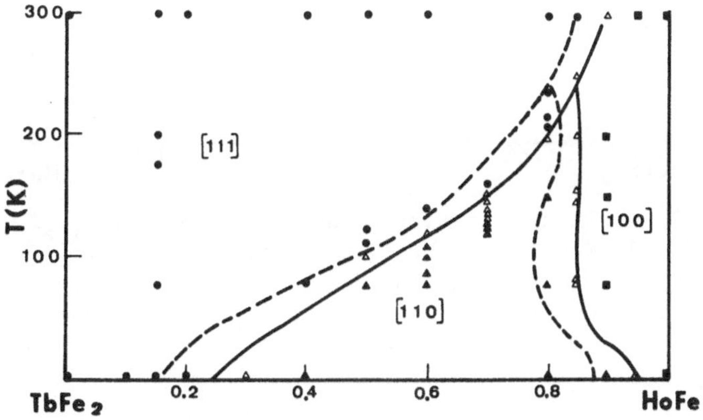

FIG. 2. Spin orientation diagram in the $Ho_xTb_{1-x}Fe_2$ system. Filled circles , filled triangles, and filled squares correspond to experimentally determined spectra characteristic of the [111], [110] and [100] axes of magnetization respectively. Open triangles correspond to intermediate types of spectra. The solid lines are the experimentally determined boundaries of regions with different directions of magnetizations. The dashed lines are the theoretical boundaries based on the one ion model.

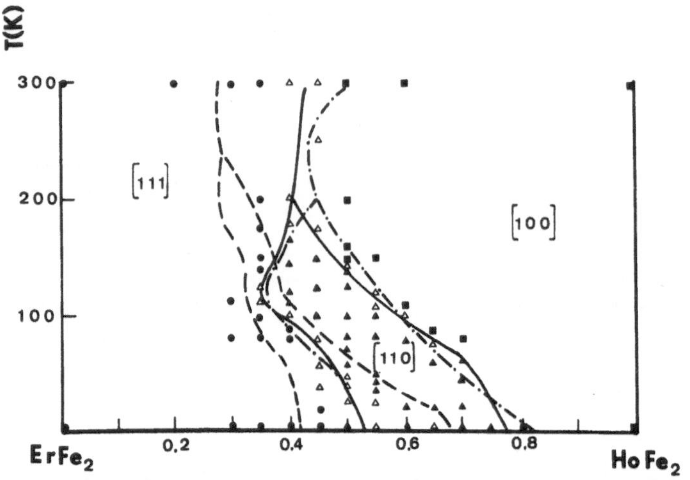

FIG. 3. Spin orientation diagram of the $Ho_xEr_{1-x}Fe_2$ system. Symbols and lines have the same meaning as in Fig. 2. The dash-dot line represents the theoretical boundaries taking account of the additional contributions to the anisotropy.

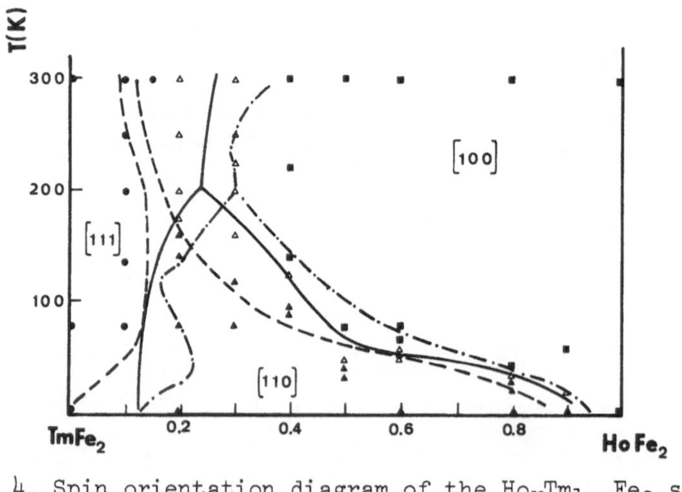

FIG. 4. Spin orientation diagram of the $Ho_xTm_{1-x}Fe_2$ system. For the meaning of symbols see Figs. 2 and 3.

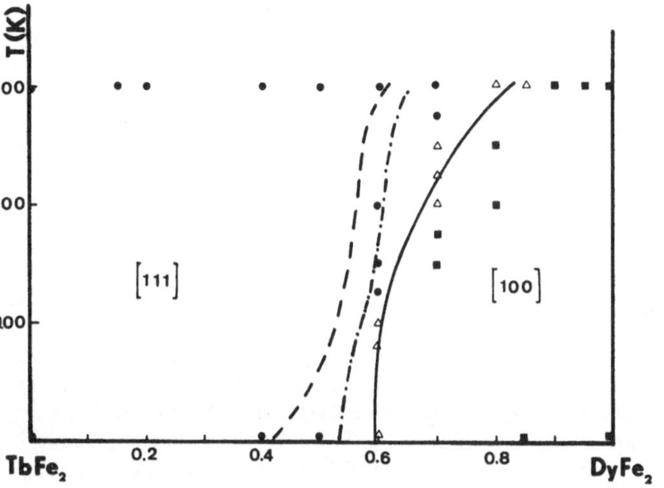

FIG. 5. Spin orientation diagrams of the $Dy_xTb_{1-x}Fe_2$ system. For the meaning of symbols see Figs. 2 and 3.

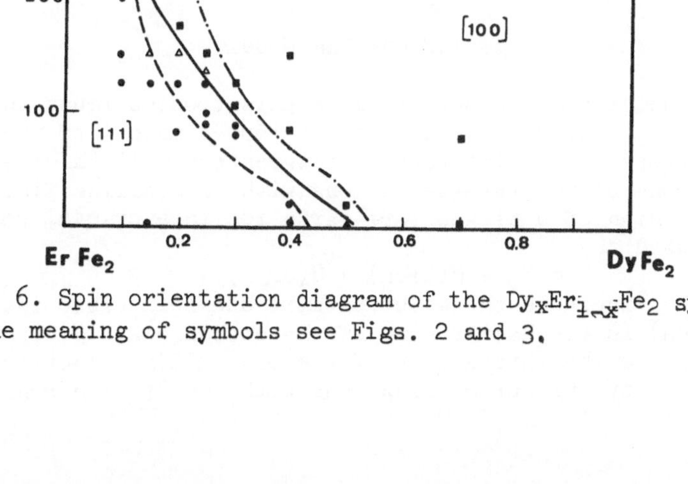

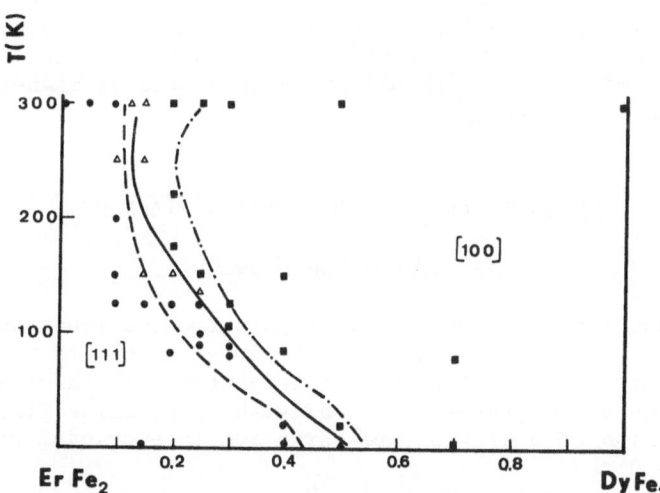

FIG. 6. Spin orientation diagram of the $Dy_xEr_{1-x}Fe_2$ system. For the meaning of symbols see Figs. 2 and 3.

examined by metallographic and electron microprobe measurements, showing that they contained less than 5% of foreign phases and that the homogeneity with respect to the distribution of the rare-earth component was better than 2%. Mossbauer effect measurements were carried out in absorbers at various temperatures between 4.2 and 300 K. The temperature stability and the accuracy of its measurement are estimated as 1 K. The experimental results are summarized in spin orientation diagrams. In these diagrams the (x,T) plane is divided into two or three regions. In each of these regions the magnetization is along one of the major cubic crystal axes. The transitions between the regions are not sharp. Spectra belonging to these regions of transition cannot be ascribed to one of the three types shown in Fig. 1 and thus the direction of easy magnetization cannot be defined. The spin orientation diagrams of the Ho containing systems $Ho_xR_{1-x}Fe_2$ (R = Tb, Er and Tm) are shown in Figs. 2 -4. In these diagrams there exists at low temperatures a region with $\vec{n}$ parallel to the [110] direction, even though at the binary limits $\vec{n}$ is parallel either to the [100] direction (x = 1) or to the [111] direction (x = 0) throughout the whole temperature range. The spin orientation diagrams of the $Dy_xR_{1-x}Fe_2$ (R = Tb, Er) systems are shown in Figs. 5 and 6. In these diagrams, only two regions appear with $\vec{n}$ either along the [100] direction for high values of x, or along the [111] direction for lower values of x. In Figs. 2 to 6 the experimentally determined boundaries (transition type spectra) of regions with different easy axes of magnetization, appear as solid lines.

THEORETICAL SPIN ORIENTATION DIAGRAMS

The One Ion Model

The high magnetocrystalline anisotropies found in rare-earth containing compounds is attributed mainly to the anisotropy of the interaction between the well shielded 4f electrons of the rare-earth ions with crystalline field. The Hamiltonian of a single rare-earth ion in a crystal can be expressed as

$$\mathcal{H} = \mathcal{H}_0 + \mathcal{H}(\text{exch}) + \mathcal{H}(\text{crys}) + \qquad (1)$$

where $\mathcal{H}_0$ includes the electrostatic and spin orbit interaction, $\mathcal{H}(\text{exch})$ is the Hamiltonian of the exchange and $\mathcal{H}(\text{crys})$ represents the interaction of the crystalline electric field produced by the surrounding ions with the 4f electrons in the partially filled shell. H_0 is isotropic and the anisotropic part of the Hamiltonian is H(exch) + H(crys). In the rare-earth (R) transition metal (M) intermetallic compounds it is usually assumed that the R-R interactions are small compared to the R-M interactions and can be neglected (5). In the present case this assumption is supported by results of Mossbauer effect measurements on ^{169}Tm in $TmFe_2$ (6) and on ^{161}Dy in $DyFe_2$ (7). The temperature dependences of the hyperfine fields acting on the rare-earth nuclei in these compounds were fitted by Brillouin functions with $\vec{J}$ appropriate to the respective rare-earth ion and exchange fields proportional to the hyperfine fields acting on the iron nuclei at the various temperatures. Furthermore the ordering temperatures of all RFe_2 compounds are much higher than the respective ordering temperatures of the isostructural RCo_2 and RNi_2 compounds, indicating again that the R-R interactions are much smaller than the R-Fe and Fe-Fe interactions. Thus the total anisotropy part of the Hamiltonian per unit volume may be obtained by summing over all rare-earth ions

$$H(\text{anis}) = \Sigma_R\{ H(\text{exch})+ H(\text{crys})\} = N\{ H(\text{exch})+ H(\text{crys})\} \quad (2)$$

where N is the number or rare-earth ions per unit volume. Within a single J manifold H(exch) can be expressed as

$$H(\text{exch}) = 2(g_J - 1)\beta_B H_{exch} \vec{J}\cdot\vec{n} \quad (3)$$

where H_{exch} is the exchange field acting on the rare-earth ion and is parallel to the easy direction of magnetization $\vec{n}$. H_{exch} is assumed to be isotropic, and to be independent of the rare-earth ion involved. It is also assumed to have the same temperature dependence as H_{eff} acting on the iron nuclei. This temperature dependence was approximated by the expression

$$H_{exch}(T) = H_{exch}(0)(1 - 0.1T/300) \quad (4)$$

for $4.2<T<300$ K. This expression is derived from Mossbauer and neutron diffraction measurements (7,8).

The Hamiltonian of the cubic crystal field interaction is written as

$$H(\text{crys}) = V_4 + V_6 \quad (5)$$

$$V_4 = A_4(1-\sigma_4) \langle r^4\rangle\langle J||\beta||J\rangle(O_4^0 + 5\ O_4^4)$$

$$V_6 = A_6(1-\sigma_6) \langle r^6\rangle\langle J||\gamma||J\rangle(O_6^0 - 21\ O_6^4)$$

with the quantization axis pointing in the [100] direction. The O_n^m are the operator equivalents, β and γ the reduced matrix elements tabulated by Hutchings (9). The values of

$\langle r^4 \rangle$ used were those tabulated for the different rare-earth ions by Freeman and Watson (10). A_4 and A_6, the crystal field parameters are assumed to be independent of the rare-earth ion involved. The values of σ_4 and σ_6, the shielding parameters have also been calculated by Freeman and Watson (11). The eigenvalues ε_i of the Hamiltonian (Eqs. 3 and 5) of the binary compounds RFe_2 were calculated for the three possible directions ($\vec{n}_j$) of the easy magnetization ([100], [110] and [111]) and for various values of the parameters A_4, A_6 and $H_{exch}(0)$. These parameters are assumed to have the same values for all the binary Laves compounds of the investigated rare-earths.

The magnetocrystalline free-energy per unit volue $F_R(\vec{n}_j,T)$ of the binary compound RFe_2 is given by

$$F_R(\vec{n}_j,T) = - k\ T\ \ln Z \tag{6}$$

where $Z(T,n_j)$ is the partition function

$$Z(T,\vec{n}_j) = \sum_{i=1}^{j} \exp(-\varepsilon_i/kT) \tag{7}$$

For the ternary compounds $R^1_x\ R^2_{1-x}\ Fe_2$ the magnetocrystalline free energy can be expressed by

$$F(x,\vec{n}_j,T) = x\ F_{R^1}(\vec{n}_j,T)+(1-x)F_{R^2}(\vec{n}_j,T) \tag{8}$$

The easy direction of magnetization of a given compound at a given temperature is the direction ($\vec{n}_j$) for which expression (8) has the lowest value. This procedure was repeated for various values of x and T and used to construct theoretical spin orientation diagrams. These diagrams still depend on the values chosen for the three parameters A_4, A_6 and $\beta_B H_{exch}(0)$.

Comparison with the experimental spin orientation diagrams (Figs. 2-6) indicates that A_6 cannot be neglected and that the ratio $A_6/A_4 = -0.038 \pm 0.003\ a_o^{-2}$ fits best the experimental data. The theoretical spin orientation diagrams of the Ho containing compounds are sensitive to the ratio A_6/A_4, but not to the exact values of A_4 and $\beta_B H_{exch}$. The experimental results were fitted with values of A_4 between 10 and 50 (K/a_o^4) and values of $\beta_B H_{exch}$ between -110 and -160 K. The values of $\beta_B H_{exch}$ are in agreement with those derived from other Mossbauer effect measurements on similar compounds (7,12). The spin orientation diagrams of Dy containing systems are not sensitive even to values of A_6/A_4 and indeed in these diagrams no region with $\vec{n}$ parallel to

[110] is found. The theoretical spin orientation diagrams with $A_4 = 10(K/a_o^4)$, $A_6/A_4 = -0.04\ a_o^4$ and $\beta_B H_{exch}(0) = -150$ K are shown in Figs. 2-6 (the dashed lines).

Additional Contributions to the Magnetic Anisotropy

Though the above outlined theory reproduces the main features of the experimental spin orientation diagrams, several systematic deviations subsist. The experimental real extension of the [111] regions seems in all instances to be larger than that predicted by the single ion model. This feature is particularly salient in the $Ho_xTb_{1-x}Fe_2$ system, in which the theoretical boundaries of the various regions are laterally shifted with respect to the experimentally determined boundaries, supposed to lie in the middle of the transition regions.

The shift between the experimental and theoretical boundaries in the spin orientation diagrams may be due to several factors. The main factor is probably neglect in the theoretical calculations of an anisotropy term contributed by the Fe-Fe interaction, which favours the [111] direction (as demonstrated by the fact that in YFe_2 (4) and $ZrFe_2$ (13) the easy magnetization is in the [111] direction). Disagreements between the theoretical and experimental spin orientation diagrams may also be caused by the uncertainty in the values of some of the parameters ($\langle r^n \rangle$, antishielding factors) used in the calculations, and by the anisotropy of the exchange interactions. Finally, the disagreement may be due to the neglect of other non-cubic anisotropy terms, such as dipolar fields or distortions, that were not taken into account in Eq.(1). The effect of distortions in particular may be non-negligible as shown recently by the giant magnetostrictive effects observed in binary RFe_2 compounds (14). Efforts have been also made in order to check whether mixing of the higher ionic J levels into the ground level has any effect on the shape of the theoretical spin orientation diagrams. Calculations taking into account such mixing were carried out and have shown that this mixing has an absolutely negligible effect on the theoretical diagrams.

In order to improve the fit between the theoretical and experimental spin orientation diagrams, it was assumed that the disagreement between them is produced only by the anisotropy of the Fe-Fe interaction, which would be the same for all RFe_2 compounds: the additional free energies in the [100],

[110] and [111] directions at various temperatures which are needed in order to get a perfect overlap between the theoretical and experimental spin orientation diagram in the $Ho_xTb_{1-x}Fe_2$ system have been calculated. The values of these additional free energies, $F_{add}(\vec{n}_j,T)$, obtained for this system, were then used to obtain corrected spin orientation diagrams in all other systems. (Actually, only the differences, $F_{add}(\vec{n}_2,T) - F_{add}(\vec{n}_1,T)$ and $F_{add}(\vec{n}_3,T) - F_{add}(\vec{n}_1,T)$ can be determined by the fitting of the theoretical to the experimental spin orientation diagrams of the $Ho_xTb_{1-x}Fe_2$ system and only these differences are relevant in order to correct theoretical spin orientation diagrams).

The following relations between the free energies at a point (x_1,T) lying on the boundary between the [111] and the [100] regions in the spin orientation diagram of the $Ho_xTb_{1-x}Fe_2$ system must hold to give a perfect fit between the experimental and theoretical orientation diagrams:

$$x_1F_{Ho}(\vec{n}_3,T) + (1-x_1)\, F_{Tb}(\vec{n}_3,T) + F_{add}(\vec{n}_3,T) = \\ = x_1F_{Ho}(\vec{n}_2,T) + (1-x_1)\, F_{Tb}(\vec{n}_2,T) + F_{add}(\vec{n}_2,T) \tag{9}$$

Similarly for a transition point (x_2,T) between the [110] and the [100] regions, the relation

$$x_2F_{Ho}(\vec{n}_2,T) + (1-x_2)\, F_{Tb}(\vec{n}_2,T) + F_{add}(\vec{n}_2,T) = \\ = x_2F_{Ho}(\vec{n}_1,T) + (1-x_2)\, F_{Tb}(\vec{n}_1,T) + F_{add}(\vec{n}_1,T) \tag{10}$$

must hold, and finally for a transition point (x_3,T) between the [111] and [100] region

$$x_3F_{Ho}(\vec{n}_3,T) + (1-x_3)\, F_{Tb}(\vec{n}_3,T) + F_{add}(\vec{n}_3,T) = \\ = x_3F_{Ho}(\vec{n}_1,T) + (1-x_3)\, F_{Tb}(\vec{n}_1,T) + F_{add}(\vec{n}_1,T) \tag{11}$$

must hold. By solving equations (9) and (10) the values of $F'_{add}(\vec{n}_2,T) = F_{add}(\vec{n}_2,T) - F_{add}(\vec{n}_1,T)$ and $F'_{add}(\vec{n}_3,T) =$ $= F_{add}(\vec{n}_3,T) - F_{add}(\vec{n}_1,T)$ were determined as a function of T for temperatures between 4 K and the temperature corresponding to the triple point in the spin orientation diagram. Using eq. (11) the values of $F'_{add}(\vec{n}_3,T)$ were determined for temperature between the temperature corresponding to the triple point, and 300 K. The values used for $F_{Ho}(\vec{n}_j,T)$ and $F_{Tb}(\vec{n}_j,T)$ were those determined previously, on basis of the single ion model assuming $A_4 = 10\text{ K} / a_o^4$, $A_6/A_4 = -0.038\ a_o^{-2}$ and $\beta_B H_{exch} = -150$ K.

Assuming the additional anisotropy energy to be independent of the rare-earth ions involved, the values of $F'_{add}(\vec{n}_2,T)$ and $F'_{add}(\vec{n}_3,T)$ found for the $Ho_{1-x}Tb_xFe_2$ system were used in all other ternary systems to obtain corrected spin orientation diagrams. The agreement between the experimental spin orientation diagrams and the theoretical diagrams is improved in all cases, showing that the anisotropy of the Fe-Fe exchange interaction (independent of R) should indeed not be neglected.

A very good agreement is obtained in the $Ho_xEr_{1-x}Fe_2$ system (fig. 3). In the $Ho_xTm_{1-x}Fe_2$ (fig. 4) and the $Dy_xTb_{1-x}Fe_2$ (fig. 5) systems the agreement between theory and experimental data is improved though it is not perfect. In the $Dy_xEr_{1-x}Fe_2$ (fig. 6) system the corrected theoretical boundary line is to the right of the experimental boundary line, whereas, the non-corrected theoretical curve lies to the left of the experimental line.

Efforts were made to improve the fit between the experimental and theoretical spin orientation diagrams by following the above procedure for determining $F'_{add}(\vec{n}_j,T)$ using values of A_4, A_6/A_4 and $\beta_B H_{exch}$ different from those mentioned above. Theoretical spin orientation diagrams were calculated for values of $10\ K/a_o^4 \leq A_4 \leq 50\ K/a_o^4$ and $-150\ K \leq \beta_B H_{exch} \leq -110\ K$ and various values of A_6/A_4. The theoretical diagrams (including $F'_{add}(\vec{n}_j,T)$ corrections) were found to be insensitive to the values of A_4 and $\beta_B H_{exch}$. The best agreement was again obtained for $A_6/A_4 = -0.038 \pm 0.003\ a_o^{-2}$.

The persisting disagreemets between the theoretical and experimental spin orientation diagrams for some of the systems indicate that $F'_{add}(\vec{n}_j,T)$ is not entirely due to the Fe-Fe interaction, but depends also on the rare-earth ion involved. The anisotropy of the exchange interaction acting on the rare-earth ion which has been neglected, is probable the main source of the remaining disagreements.

Transition Regions

The theoretical model predicts a first-order phase transition between the regions of different spin orientation. This prediction is in apparent contradiction with the presence of the transition regions in the spin-orientation diagram. The appearance of the transition regions may be due to the following factors:

1. The inhomogeneity of the samples. This however cannot be the main factor, as some of the experimental spectra (e.g. transition spectra) cannot be interpreted as a superposition of two spectra corresponding to magnetizations along two different major crystalline directions.
2. The existence of additional , noncubic anisotropic terms due to distortions, dipolar fields, and anisotropic exchange that have been neglected in Eq.(1), As a result of these terms, the magnetization may not coincide with the direction of one of the major crystalline axes, and in the transition region $\vec{n}$ may deviate significantly from these axes.

CONCLUSIONS

The present experimental data concerning the direction of magnetization of the $R^1{}_xR^2{}_{1-x}Fe_2$ compounds were described in terms of (x,T) spin orientation diagrams. Theoretical spin-orientation diagrams were calculated assuming that the magnetic crystalline anisotropy is due to the anisotropy of the interaction between the 4f electrons of the rare-earth ions with the crystalline fields. These theoretical calculations reproduced the general features of the experimental results though small discrepancies remained. Adding to the calculated free energy of the crystals a contribution which originates mainly in the Fe-Fe interaction and is independent of the rare-earth ions involved, improved the agreement between the theoretical and experimental spin orientation diagrams. It seems that the remaining discrepancies between theory and experiment are a result of the neglect of non cubic anisotopic terms due to crystalline distortion, dipolar fields and especially the anisotropy of the Fe-R exchange interactions. These last terms are probably also responsible for the existence of the transition regions.

REFERENCES

1. A large number of references are to be found for instance in: L.M. Levinson, M. Luban, and S. Shtrikman, Phys. Rev., 187, 715 (1969) or G. Gorodetsky, B. Lüthi, and T. J. Moran, Intern. J. Magnetism, 1, 295 (1971).

2. U. Atzmony, K. Hardy, and J.C. Walker, J. Phys. (Paris), 32 C1-920 (1971).

3. U. Atzmony, M.P. Dariel, E.R. Bauminger, D. Lebenbaum, I. Nowik, and S. Ofer, to be published.

4. G.J. Bowden, D. St. P. Bunbury, A.P. Guimaraes, and R.E. Snyder, J. Phys. C. Proc. Phys. Soc., London, 2, 1367 (1968).

5. K.H.J. Buschow, Phys. Stat. Sol. (A) 7, 199 (1971).

6. R.L. Cohen, Phys. Rev., 134, A94 (1964).

7. E. Segal, Ph.D. Thesis, Hebrew University, 1967 (unpublished).

8. G. Will and M.O. Bargouth, Phys. Kondens. Materie 13, 137 (1971).

9. M.T. Hutchings, in Solid State Physics, edited by H. Ehrenreich, F. Seitz, and D. Turnbull (Academic, New York, 1966), Vol. 16, 0. 277.

10. A.J. Freeman and R.E. Watson, Phys. Rev. 127, 2058 (1962).

11. A.J. Freeman and R.E. Watson, Phys. Rev. 139, A/606 (1965).

12. Ref. 6. as quoted by S. Ofer, I. Nowik and S.G. Cohen, in "Chemical Applications of Mossbauer Spectroscopy" edited by I.V.I. Goldanski and R.H. Herber, (Academic, New York, 1968) p. 471.

13. G.K. Wertheim, V. Jaccarino, and J.H. Wernick, Phys. Rev. 135, A151 (1964).

14. A.E. Clark and H.S. Belson, Phys. Rev. B, 5, 3642 (1972).

^{57}FE AND ^{99}RU MÖSSBAUER STUDIES OF THE DEFECT STRUCTURE AND MAGNETIC PROPERTIES OF OXIDE PHASES

N.N. Greenwood

Department of Inorganic and Structural Chemistry

University of Leeds, Leeds LS2 9JT, England

The extensive use of Mössbauer spectroscopy to study oxide phases has recently been fully reviewed[1,2] and little point would be served in giving another general survey. Instead, I would like to discuss in some depth the contributions which Mössbauer spectroscopy has made to two key systems: (a) the defect structure of nonstoichiometric iron monoxide $Fe_{1-x}O$ and (b) the magnetic properties of ternary and quaternary ruthenium oxides.

In the case of iron monoxide, Mössbauer spectroscopy should ideally enable us (i) to differentiate between Fe^{2+} and Fe^{3+}, (ii) to differentiate between octahedral and tetrahedral sites, (iii) to determine whether Fe^{2+} and Fe^{3+} exist as such on the sub-microsecond time scale or whether electron hopping wipes out this distinction as it does in Fe_3O_4, (iv) to indicate distortions from ideal cubic site symmetry, (v) to determine the Néel temperature as a function of varying composition, (vi) to determine the rates at which $Fe_{1-x}O$ disproportionates into other protoxides of iron and finally into Fe and Fe_3O_4, (vii) to determine the rate of diffusion of Fe-ions through the lattice at high temperatures by means of line broadening studies. Because of the importance of the FeO system several aspects have previously been studied[1] but it is only very recently that a comprehensive investigation has been undertaken[3,4].

In the case of the ternary and quaternary ruthenium oxides preliminary results[5] have indicated the value of the ruthenium-99 resonance in probing the magnetic structure of these phases and the work is currently being rapidly extended.[6,7] A further important aspect of this work is the comparison between iron and ruthenium which are neighbouring elements in the same group of the Periodic Table.

The Wüstite Phase.- The oxidation of iron in the presence of a limited amount of oxygen usually results in a phase of composition between $Fe_{0.95}O$ and $Fe_{0.85}O$, the wüstite phase. All attempts to prepare stoichiometric FeO (except perhaps under very high-pressure conditions[8]) leads to precipitation of iron, and even the phase $Fe_{1-x}O$ is unstable below 570°C, tending to disproportionate finally into iron and magnetite.

$$Fe + \tfrac{1}{2}O_2 \rightarrow [FeO] \rightarrow xFe + Fe_{1-x}O$$
$$4Fe_{1-x}O \rightarrow (1-4x)Fe + Fe_3O_4$$

This wüstite phase itself embraces three regions (I, II, and III) delineated by second-order phase transitions, and quenched specimens at room temperature show the presence of three metastable phases P, P' and P''. These phase relations are shown in Figure 1.

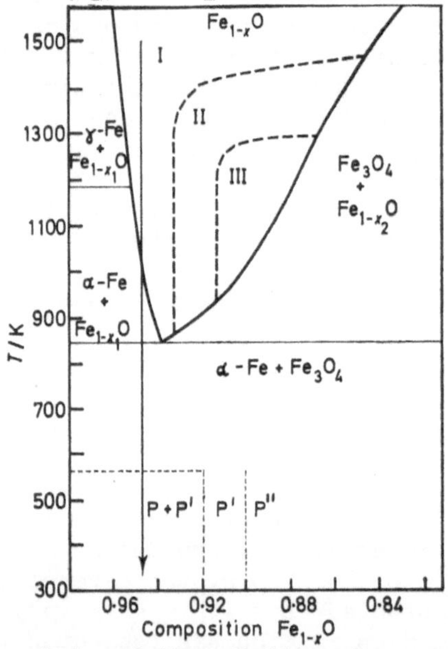

Fig.1.

A Mössbauer study of the wüstite phase under equilibrium conditions is fraught with difficulties: the high temperatures involve a diminution in recoil-free fraction, the oxygen partial pressure must be precisely maintained at various values in the region of 10^{-20} atm, the highly reactive oxide must be supported on an inert, vibration-free mount etc. However, with appropriate instrumentation[3] it is possible to obtain good spectra and these give valuable information about the diffusion processes occurring at high temperature. If the resonant ions are not bound rigidly in the lattice but migrate by hopping from site to site, then the resonance line becomes broadened by an amount $\Delta\Gamma$ when the time of residence on one site τ becomes comparable with the half-life of the ^{57}Fe excited state (99.7 ns). As shown by Singwi and Sjölander,

$$\Delta\Gamma = 2\hbar\tau^{-1}$$

Typical spectra are given in Fig.2a which shows that, for $Fe_{0.91}O$, diffusion broadening increases the linewidth more than fourfold at 800°C and more than 15-fold at 900°C. Variation with composition is shown in Fig.2b. The results lead to activation energies which are virtually independent

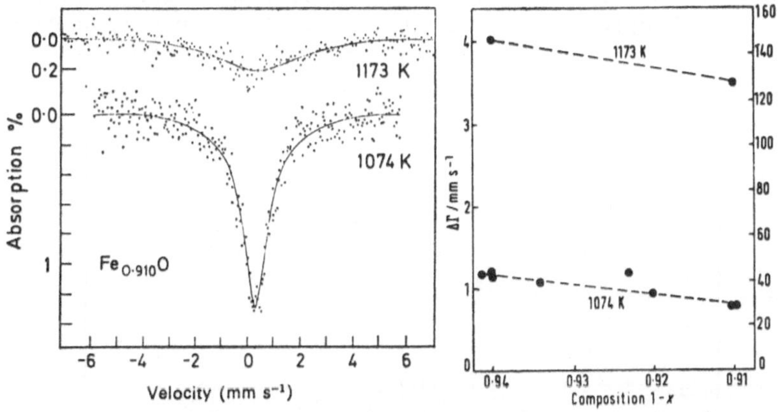

Fig.2 (a) Diffusion-broadened spectra of $Fe_{0.91}O$ at two temperatures. (b) Influence of composition and temperature on diffusion broadening $\Delta\Gamma$ and jump frequency τ^{-1}

of concentration, e.g. 140 ± 20 kJ mol^{-1} for $Fe_{0.94}O$ and 135 ± 20 kJ mol for $Fe_{0.91}O$. This is consistent with defect clustering as discussed later. The experimentally determined jump frequency can also be used to calculate the diffusion coefficient D by means of the relation

$$D = (r_0^2 f \tau^{-1})/6$$

where r_0 is the jump distance and f is the Bardeen-Herring correlation coefficient which allows for the possibility that a cation might jump back into the site from which it came, thus contributing to the hopping frequency but not to the diffusion process. Values of D so obtained are in the range (0.5-2) x $10^{-8} cm^2 s^{-1}$ and are similar to values obtained by radioactive tracer techniques. The Mössbauer technique however is the more reliable for determining jump frequencies because it eliminates errors due to scaling, grain boundaries, and pre-equilibration effects.

Quenched Samples of $Fe_{1-x}O$.– Below 570° wüstite rapidly disproportionates into iron and magnetite (Fig.1) but specimens of $Fe_{1-x}O$ can be obtained at lower temperatures providing they are quenched sufficiently rapidly through the sensitive region. The room temperature Mössbauer spectra of nonstoichiometric iron monoxides depend not only on the overall composition but also on the temperature from which they are quenched and the rate of quenching. Extensive preliminary experiments established the most favourable conditions[3] and enabled a study of $Fe_{1-x}O$ at and below room temperature where disproportionation is kinetically hindered. Typical spectra covering the complete composition range

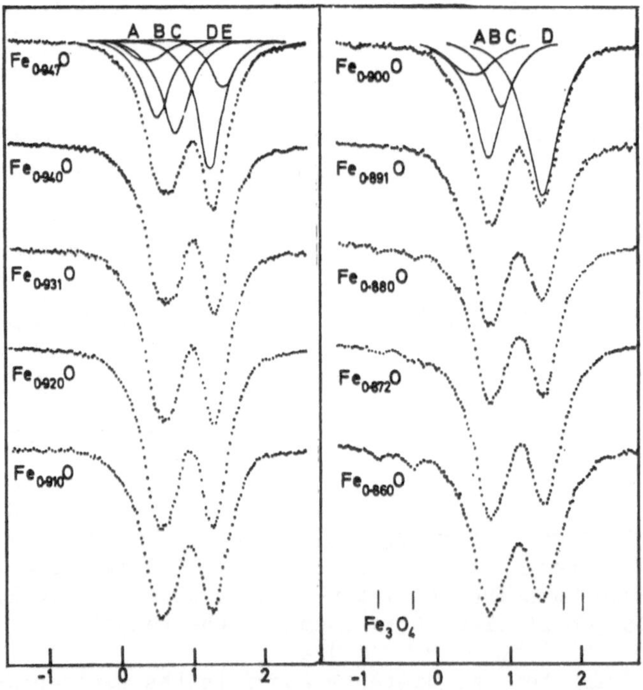

Fig.3 Room-temperature spectra of $Fe_{1-x}O$ quenched from 1250°C into water, showing the progressive change of the resonance profile with composition.

are shown in Fig.3. Detailed analysis of these spectra in which 10 different models were examined yield the following results. The spectrum of $Fe_{0.947}O$ is best represented as the sum of 5 Lorentzians; curve A is one half of an Fe^{3+} quadrupole split doublet the other half of which is under curve C; B-E and C-D are quadrupole split doublets from Fe^{2+}. On this model normal values for chemical shifts and quadrupole splittings are obtained and the intensity of the Fe^{3+} resonance correlates well with the known composition $Fe^{2+}_{0.841}Fe^{3+}_{0.106}O$. This also implies that Fe^{3+} and Fe^{2+} give separate resonances and that electron hopping is distinctly slower than the Mössbauer timescale (10^{-7}s). As the composition deviates further from stoichiometry, the amount of Fe^{3+} builds up rapidly and this is seen as an increase in intensity of the low energy limb of the spectrum. However, below $Fe_{0.91}O$ it is no longer possible to resolve the envelope into the sum of a small number of Lorentzians because of the large number of environments available for Fe^{3+} and Fe^{2+}. The extent of the stoichiometric disaster afflicting the $Fe_{1-x}O$ phase can be gauged from the following tabulation:

Compound	FeO	←	$Fe_{1-x}O$	→	Fe_3O_4	Fe_2O_3
Ratio Fe/O	1.00	0.95	0.90	0.85	0.75	0.67
Fe^{3+}/Fe(tot)%	0.0	10.5	22.2	35.3	66.7	100

Histograms showing the proportion of Fe^{2+} ions having 0-9 neighbouring defects (Fe^{3+} or cation vacancy) reinforce these conclusions.[3] The form of such histograms depends on the structural model adopted for the lattice defects present in the nonstoichiometric phase. The simplest model (1), based on the presence of isolated Schottky type defects in the cation lattice of the predominantly sodium-chloride-like structure, is unlikely because of the high defect concentration and the coulombic attraction between Fe^{3+} and the virtual negative charge on the cation vacancy, $\square$. Simple binary clusters are envisaged in model 2.

Model 1. $[Fe^{2+}_{1-3x}]_{oct}[Fe^{3+}_{2x}]_{oct}[\square_x]_{oct}O$

Model 2. $\{[Fe^{3+}]_{oct}[\square]_{oct}\}$

Model 3. $\{(Fe^{3+})_{tet}[\square_4]_{oct}\}^{5-}$ plus $5[Fe^{3+}]_{oct}$ nearby

Model 4. $\{(Fe^{3+}_4)_{tet}[\square_{13}]_{oct}\}^{14-}$ plus $14[Fe^{3+}]_{oct}$ nearby

The well known preference of Fe^{3+} for tetrahedral sites further increases the lattice energy but is only sterically possible if the four neighbouring tetrahedral sites are vacant (model 3): there is good evidence that some of the Fe^{3+} is in this type of cluster from X-ray and neutron diffraction experiments and the Mössbauer spectra of samples in the range $Fe_{0.95}O$ to $Fe_{0.91}O$ can be interpreted on this basis. However as the concentration of Fe^{3+} and vacant cation sites increases beyond this point, further clustering occurs to give the cubic Koch-Cohen cluster (model 4) and analysis in terms of two predominant types of Fe^{2+} environment is no longer possible.

Magnetic spectra of quenched $Fe_{1-x}O$ also show separate resonances from Fe^{3+} and Fe^{2+} and a range of hyperfine fields. (see Fig.4). The hyperfine field at Fe^{3+} covers the range 480 ± 10 kOe and the field at Fe^{2+} spans the range 340 ± 20 kOe. Note also the increase in strength of the Fe^{3+} resonance as the composition moves further away from stoichiometry. The Néel temperature is essentially independent of composition, being 196 ± 3 K for $Fe_{0.944}O$ and 198 ± 3 K for $Fe_{0.870}O$. This again argues for defect clusters rather than isolated defects.[3]

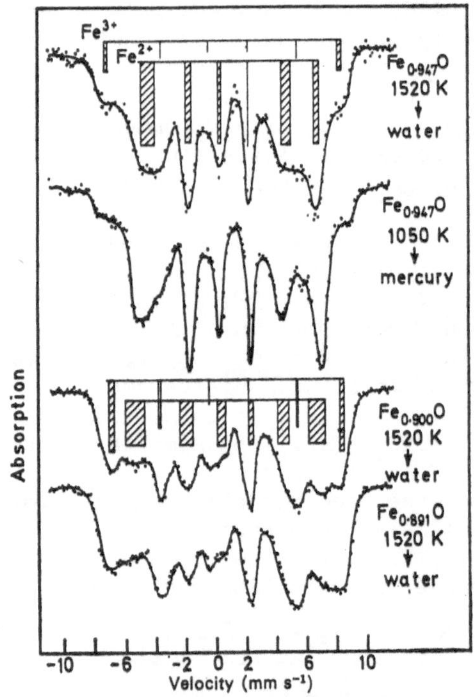

Fig.4 Mössbauer spectra of quenched samples of $Fe_{1-x}O$ at 77K.

Disproportionation of $Fe_{1-x}O$ above Room Temperature.– One of the most remarkable features of $Fe_{1-x}O$ is its very ready transformation to other phases at slightly elevated temperatures (see ref.3 for literature review). These changes can be monitored rapidly and sensitively by Mössbauer spectroscopy. Three stages can be identified:

Stage 1 (minutes at 250°C)

$$2Fe_{1-x}O \rightarrow Fe_{1-x-y}O + Fe_{1-x+y}O$$

Stage 2 (hours at 250°C)

$$(1-4z)Fe_{1-x}O \rightarrow (1-4x)Fe_{1-z}O + (x-z)Fe_3O_4$$

Stage 3 (days at 300°C)

$$4Fe_{1-z}O \rightarrow (1-4z)Fe + Fe_3O_4$$

In stage 2, when $z \rightarrow 0$ the first product approaches stoichiometric $Fe_{1.00}O$. The process is seen as a dramatic diminution in linewidth from the broad quadrupole-split envelope to a single narrow line approaching natural linewidth (see Fig.5). An idea of the amount of information

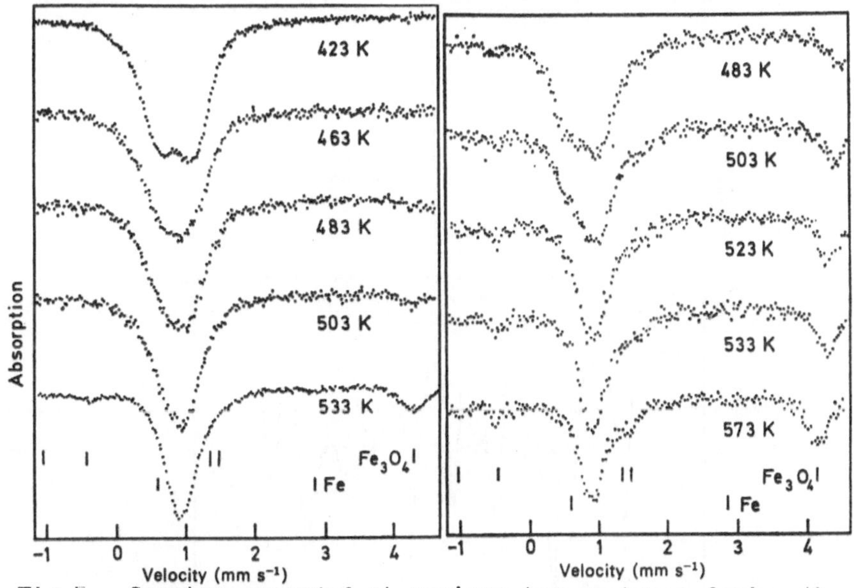

Fig.5 Spectra recorded at various temperatures during the disproportionation of $Fe_{0.94}O$ (left hand sequence) and $Fe_{0.88}O$ (right hand sequence).

obtainable from such systems can be seen from the spectra in Fig.6 which refers to a sample of $Fe_{0.94}O$ heated at 400°C for 30 min and then cooled to room temperature: the spectra of Fe and Fe_3O_4 are clearly resolved as is the narrow paramagnetic spectrum of $Fe_{0.99}O$. At 77 K this narrow line

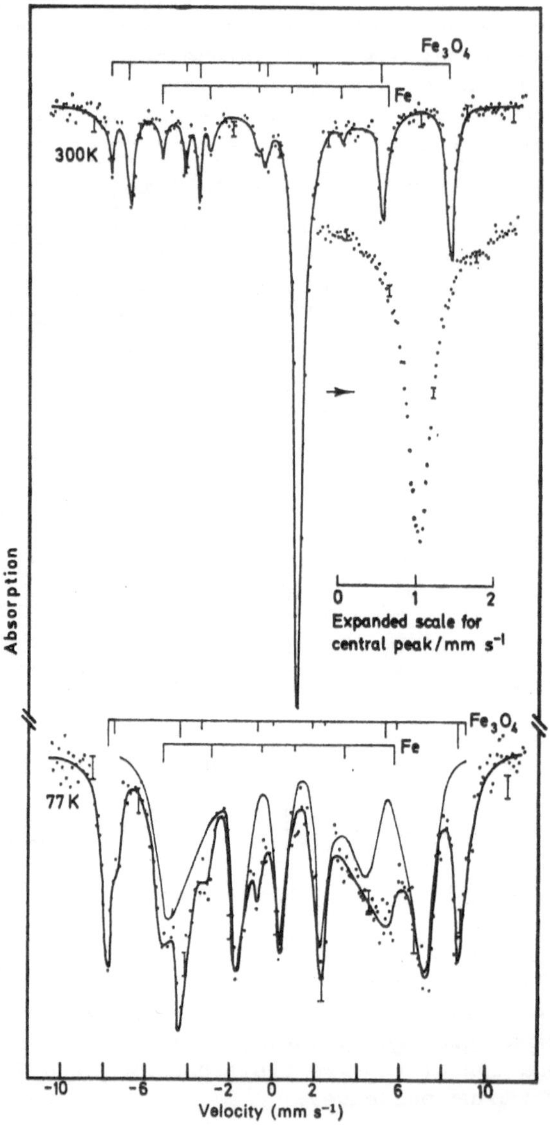

Fig.6 Spectra at 300 K and 77 K of $Fe_{0.94}O$ previously heated at 673 K for 30 min. The paramagnetic peak (Γ = 0.39 mm s^{-1}) is also shown on an expanded scale.

itself splits into a six-line magnetic spectrum (feint line) the quadrupole splitting being due to a rhombohedral distortion which occurs below the Néel temperature (197 K). The transition from the paramagnetic to the antiferromagnetically ordered state occurs rapidly (within 1 K) and there was no evidence for ordered magnetic behaviour above the Néel temperature due to intergrowths of Fe_3O_4 as had previously been suggested.

The successive stages of the disproportionation are clearly discernible in Fig.7 which is a composite plot of temperature versus time (top half) together with an indication of the time variation of the quadrupole splitting of $Fe_{1-x}O$ and the build up of Fe_3O_4 and Fe in a sample of initial composition $Fe_{0.94}O$. Initially, as the temperature

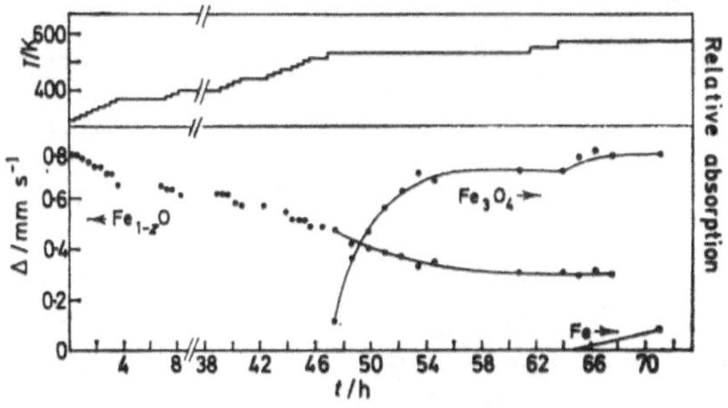

Fig.7 Changes in the spectrum of $Fe_{0.94}O$ prior to and during decomposition (see text).

is raised the quadrupole splitting decreases because of the increasing thermal population of the higher electronic levels of Fe^{2+}, but the splitting remains constant over a period of time at constant temperature. At 513 K peaks due to Fe_3O_4 appear and during a prolonged stay at 533 K these rapidly increased to a maximum; concurrently there was a steady reduction in the quadrupole splitting despite the constant temperature (Stage 2). After a further increase in the temperature to 573 K the intensity of the Fe_3O_4 peaks again increased and this was accompanied by the first appearance of Fe peaks. (Stage 3). No further narrowing of the FeO peak occurred but its intensity gradually diminished as disproportionation proceeded.

This Mössbauer investigation of nonstoichiometric iron monoxide has been presented in some detail not only because of the intrinsic importance of this system in the development of a coherent theory of the defect solid state but also because it illustrates in a particularly striking way the versatility of the Mössbauer technique and the wide range of diverse phenomena it can probe.

Ruthenium-99 Mössbauer Spectroscopy.— Ruthenium and iron are the only pair of vertically contiguous elements in the Periodic Table which can both be studied conveniently by Mössbauer spectroscopy. This, coupled with the extensive range of magnetic phenomena already investigated in iron oxide and halide systems, and the dearth of corresponding information on 4d elements, makes the use of ^{99}Ru for such studies particularly attractive. The Mössbauer systematics

TABLE Comparison of Nuclear Properties of ^{99}Ru and ^{57}Fe

Property	^{99}Ru	^{57}Fe
Parent	^{99}Rh(E.C. 16d)	Co(E.C. 270d)
E_γ/keV	90	14.4
Γ/(mm s^{-1})	0.15	0.19
I_g	$\frac{5}{2}+$	$\frac{1}{2}-$
I_e	$3/2+$	$3/2-$
Nat.ab./%	12.63	2.17
Radiation	E2 and M1, $\delta^2 = 2.7$	M1
μ_g/n.m.	-0.62	+0.0902
μ_e/n.m.	-0.29	-0.1547
Q_g/barn	$\pm$0.05	0
Q_e/barn	$\pm$0.15	0.2
$\Delta R/R$	positive	negative

for ^{99}Ru were first elucidated by Kistner and his colleagues (see ref.1 for literature review). The relevant nuclear properties are compared in the Table from which it is apparent that the line width and natural abundance of ^{99}Ru are particularly favourable, though the high energy of the γ-photon necessitates keeping both source and absorber at liquid helium temperatures and the half life of the ^{99}Rh precursor is rather short. The numerical values of the nuclear spin quantum numbers and nuclear quadrupole moments lead to an overlapping set of six lines in an electric field gradient and the resultant spectrum is usually seen as a broad, poorly resolved doublet (Fig.8).[11] The substantial admixture of E2 radiation with M1 leads to 18 allowed

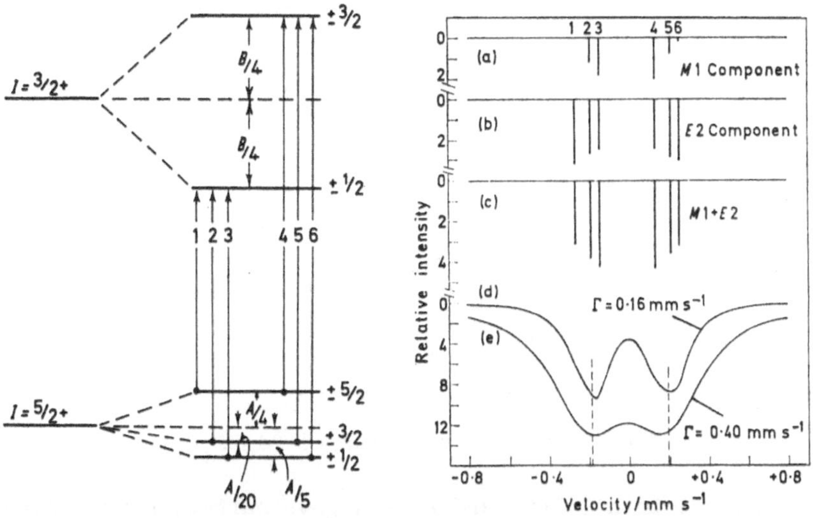

Fig.8 Energy level diagram (left) and simulated spectra (right) arising from an axially symmetric positive quadrupole interaction on ^{99}Ru.

magnetic transitions and the values of the nuclear magnetic moments lead, fortuitously, to an essentially equal spacing of the lines though these are not fully resolved in the hyperfine fields so far encountered.

Ruthenium Oxide Phases.— The ternary oxides $CaRuO_3$, and $SrRuO_3$, have slightly distorted perowskite structures and are good electrical conductors.[12] $BaRuO_3$ also has a high electronic conduction but has a 9-layer hexagonal structure. Magnetic susceptibility measurements showed that $SrRuO_3$ was ferromagnetic (T_c 160 ± 10 K), the first reported case for a second row transition element.[12] $BaRuO_3$ was described as antiferromagnetic[12] and $CaRuO_3$ was said to be antiferromagnetic with a small amount of parasitic ferromagnetism.[14]

The Mössbauer spectrum of $SrRuO_3$ at 4.2 K (Fig.9) confirms the magnetic ordering and leads to a hyperfine magnetic field of 352 ± 15 kOe consistent with 1 unpaired electron. The chemical isomer shift is -0.325 ± 0.007 mm s^{-1} relative to metallic ruthenium; this is slightly lower than normal for Ru(IV) and may indicate some oxygen deficiency.

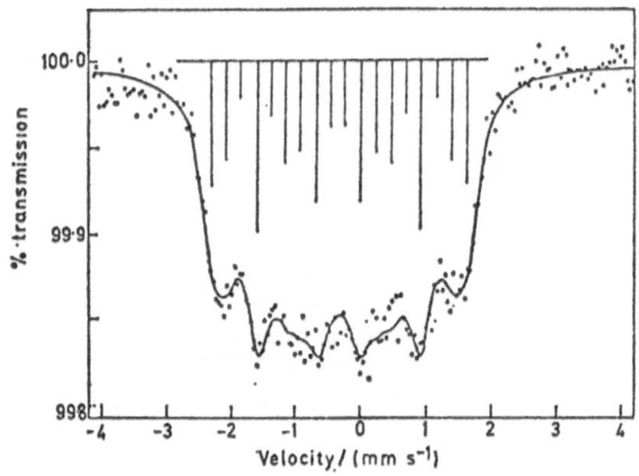

Fig.9 The ^{99}Ru Mössbauer spectrum of $SrRuO_3$ at 4.2 K.

By contrast both $CaRuO_3$ and $BaRuO_3$ gave narrow, single-line spectra with no evidence for long-range magnetic exchange interactions. Bulk susceptibility measurements down to 4.2 K confirmed that the earlier interpretations [12,14] were in error. In the 9-layer hexagonal structure of $BaRuO_3$ linear clusters of three face-sharing RuO_6 octahedra are linked by corner sharing to similar clusters. The point group symmetry of the cluster is D_{3d} and the three interacting 4d orbitals on each of the three ruthenium atoms are d_{z^2} (A_{1g} in D_{3d}), $d_{x^2-y^2}$ and d_{xy} which are degenerate (E_g). These combine to give $2A_{1g} + A_{2u} + 2E_g + E_u$ as shown in the diagram. The twelve electrons from the three ruthenium $4d^4$ ions are thus

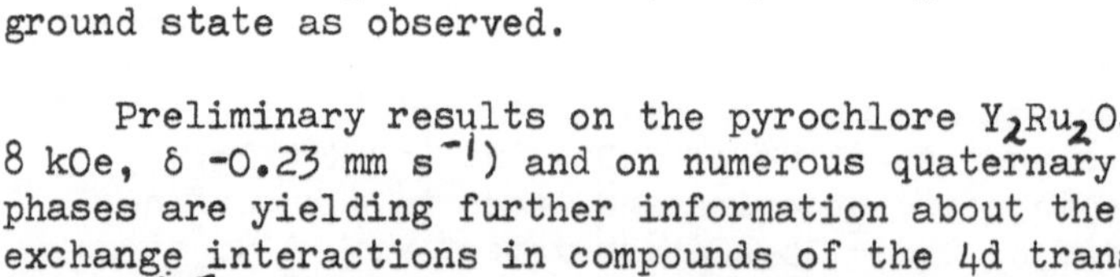

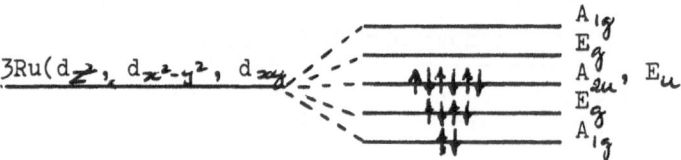

distributed to give an essentially diamagnetic cluster in the ground state as observed.

Preliminary results on the pyrochlore $Y_2Ru_2O_7$ (H 126 ± 8 kOe, δ -0.23 mm s^{-1}) and on numerous quaternary oxide phases are yielding further information about the magnetic exchange interactions in compounds of the 4d transition series [6,7]. Systems studied include $Ca_xSr_{1-x}RuO_3$, $SrRu_xIr_{1-x}O_3$, $SrRu_xMn_{1-x}O_3$ and $SrRu_xFe_{1-x}O_3$. In this latter system both Mössbauer elements can be used to probe the magnetic interactions at sites having a wide range of nearest-neighbour environments in the lattice.

References

1. N.N. Greenwood and T.C. Gibb, "Mössbauer Spectroscopy", Chapman and Hall Ltd., London, 1971.pp.xii + 659.
2. N.N. Greenwood (Senior Reporter) "Spectroscopic Properties of Inorganic and Organometallic Compounds", The Chemical Society, London Vols. 1-5, 1968-1972.
3. N.N. Greenwood and A.T. Howe, J. Chem. Soc. Dalton Trans., 1972, 110, 116, 122.
4. N.N. Greenwood and A.T. Howe, Proc. 7th International Symposium on the Reactivity of Solids, Bristol 1972, in press.
5. T.C. Gibb, R. Greatrex, N.N. Greenwood, and P. Kaspi, Chem. Commun., 1971, 319.
6. P. Kaspi, "Ruthenium-99 Mössbauer Spectroscopy", Ph.D. Thesis, University of Newcastle upon Tyne, 1972.
7. R. Greatrex, N.N. Greenwood, and K. Snowdon, unpublished results, University of Leeds, 1972.
8. T. Katsura, B. Iwasaki, S. Kimura, and S. Akimoto, J. Chem. Phys., 1967, 47, 4559.
9. K.S. Singwi and A. Sjölander, Phys. Rev., 1960, 120, 1093.
10. H. Shechter, P. Hillman, and M. Ron, J. Applied Phys., 1966, 37, 3043.
11. R. Greatrex, N.N. Greenwood, and P. Kaspi, J. Chem. Soc. (A), 1971, 1873.
12. A. Callaghan, C.W. Moeller, and R. Ward, Inorg. Chem., 1966, 5, 1572.
13. P.C. Donahue, L. Katz, and R. Ward, Inorg. Chem., 1965, 4, 306.
14. J.M. Longo, P.M. Raccah, and J.B. Goodenough, J. Appl. Phys., 1968, 39, 1327.

X-RAY PHOTOELECTRON SPECTROSCOPY AND MÖSSBAUER EFFECT, COMPLEMENTARY TECHNIQUES FOR THE STUDY OF SOLIDS

G. K. Wertheim

Bell Laboratories

Murray Hill, New Jersey 07974

X-ray photoelectron spectroscopy (XPS) is a new technique which has begun to make substantial contributions to our understanding of the electronic properties of solids. The basic principle is that of any photoelectric measurement. Photons of well-defined energy are incident on a specimen and produce photoelectrons whose kinetic energy is measured. Then, knowing the work function of the spectrometer the original electron binding energy can be directly obtained. We owe the development of this technique to Kai Siegbahn (1,2), who coined the acronym ESCA, Electron Spectroscopy for Chemical Analysis, to emphasize one of the first major applications. He and his coworkers did, however, also pioneer applications atomic, molecular and solid state physics.

Mössbauer effect (3) spectroscopy (MS) is another technique which has made substantial contributions to the study of solids during the past twelve years. The measurable is the perturbation of nuclear levels by electrostatic or magnetic hyperfine coupling or by second-order Doppler effect. Hyperfine coupling depend on the properties of the electronic wavefunctions at the nucleus and on the orbital moments of unfilled shells. The second-order Doppler shift is due to the mean square thermal velocity of the atoms. XPS and MS both provide information concerning the electrons in solids. XPS measures binding energies, Fig. 1, exchange splittings and densities of state; MS samples the electronic charge and spin density at the nucleus. We will

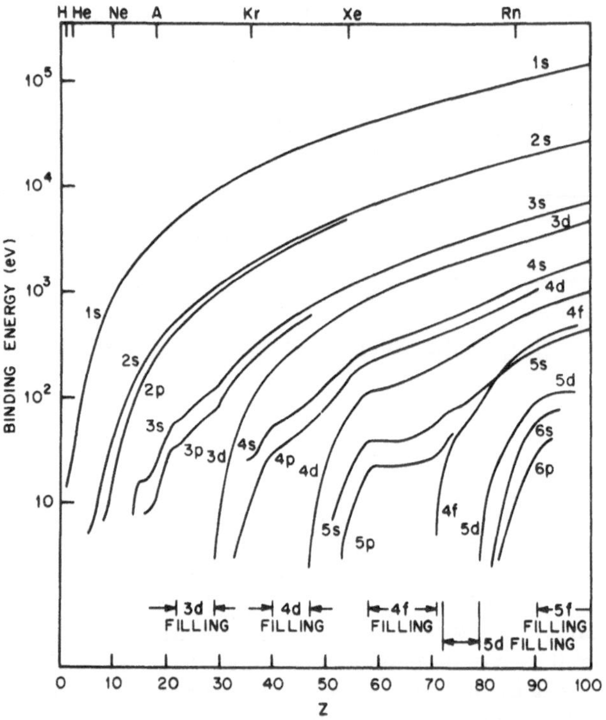

Fig. 1. Binding energies of electrons in the elements. Spin-orbit splitting has been suppressed. Chemical shifts which amount to many eV can modify the data at low E_b. (Plotted from data in ref. 1.)

examine how the information from these two techniques can be combined to deepen our understanding of the electronic properties of solids.

In MS, energy resolution of the order 10^{-9} eV is readily achieved but the measurement of electronic properties is indirect. X-ray photoemission spectroscopy, on the other hand, looks directly at the electrons with a resolution currently no better than 0.5 eV. The resolution is limited by the natural linewidth of x-rays and their unresolved spin-orbit splitting. Most of the instruments in use today employ a conventional X-ray tube with Mg or Al anode as the source of radiation, Fig. 2. Those metals were chosen because they provide the narrowest practical

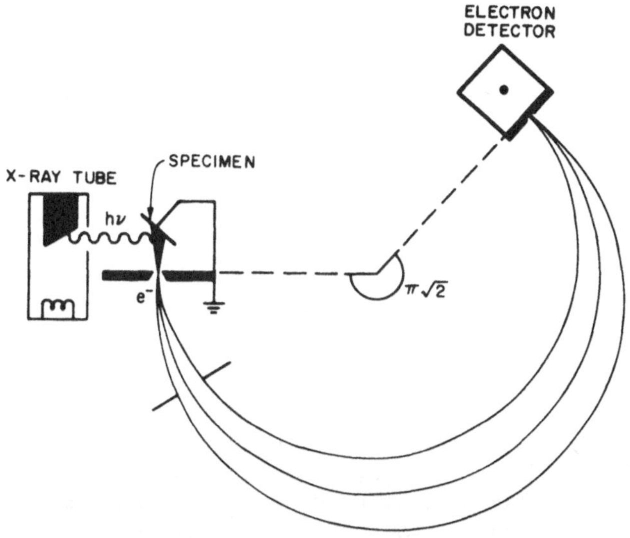

Fig. 2. X-ray photoelectron spectrometer.

X-ray lines, $MgK\alpha_{1,2}$ 1253.6 eV with a width of 0.9 eV and $AlK\alpha_{1,2}$, 1486.6 eV, 1.4 eV. Bent crystal monochromators have recently been used to improve the resolution (4) and suppress background. In the future monochromatized synchrotron radiation (5) will provide a narrow tunable source for such work. Energy analysis is usually carried out with a deflection spectrometer. Because the kinetic energy of the electrons is low, electrostatic spectrometers (6) can be used in preference to the magnetic instruments familiar from beta-ray spectroscopy. In many cases a combination of retardation followed by electrostatic analysis has proved advantageous (7).

A fundamental characteristic of these measurements is that only a thin layer of material, defined by the distance which low-energy electrons can traverse in solids without energy-loss, is actually under observation. The X-rays, of course, penetrate more deeply than the scattering length of electrons of equal energy. As a result each photoelectron line is accompanied by a tail of degraded electrons which have undergone energy loss in the solid by exciting plasmons or electronic transitions. In practice this does not seriously impair the resolution of these measurements because plasmon energy losses are usually greater than the photoelectron linewidth. The thickness of the layer from which electrons escape without energy loss is 10 to 20 Å, i.e. only a few atomic layers. (8,9,10) This makes the technique very sensitive to surface properties and eminently suitable for the study of catalysis (11) and corrosion.

After this brief description of the XPS technique we turn to examine two areas where X-ray photoelectron spectroscopy and Mössbauer effect yield related information:

1. Magnetic Hyperfine Interactions and Exchange Splitting

The magnetic properties of atoms arise from electrons in partially filled shells, e.g. 3d or 4f shells. The magnetic hyperfine splitting observed for such atoms in MS arises through the following sequence of events. (12) (We consider the spin-only case characteristic of a half-filled shell.) The electrons in incomplete shells polarize inner filled shells, including the s shells which have a finite charge density at the nucleus. This core-electron polarization produces a net spin density at the nucleus which results in a magnetic coupling between nuclear and electronic spin through the Fermi contact interaction. The Mössbauer effect thus measures the total spin density at the nucleus due to core polarization, a result which is usually expressed as an effective magnetic field.

Hartree-Fock (H-F) calculations indicate that exchange polarization produces energy differences of many electron volts between spin-up and spin-down electrons in inner s shells. In photoemission spectroscopy these electrons can be separately examined, and their exchange splitting observed. (13) This s-electron splitting is best thought of in terms of the final state of the photoemission process. An open 3d or 4f shell of some L and S couples to the spin of the remaining s-electron to give states with the same L, and $S \pm 1/2$. The result is a doublet with intensity ratio $(S+1)/S$ which is in fact observed, Fig. 3. This splitting is a single-ion property which is observable in the paramagnetic solid. A determination of the splitting of the inner s-shells together with a measurement of the hyperfine field provides a sensitive test of Hartree-Fock calculations. Some progress toward this goal has recently been achieved.

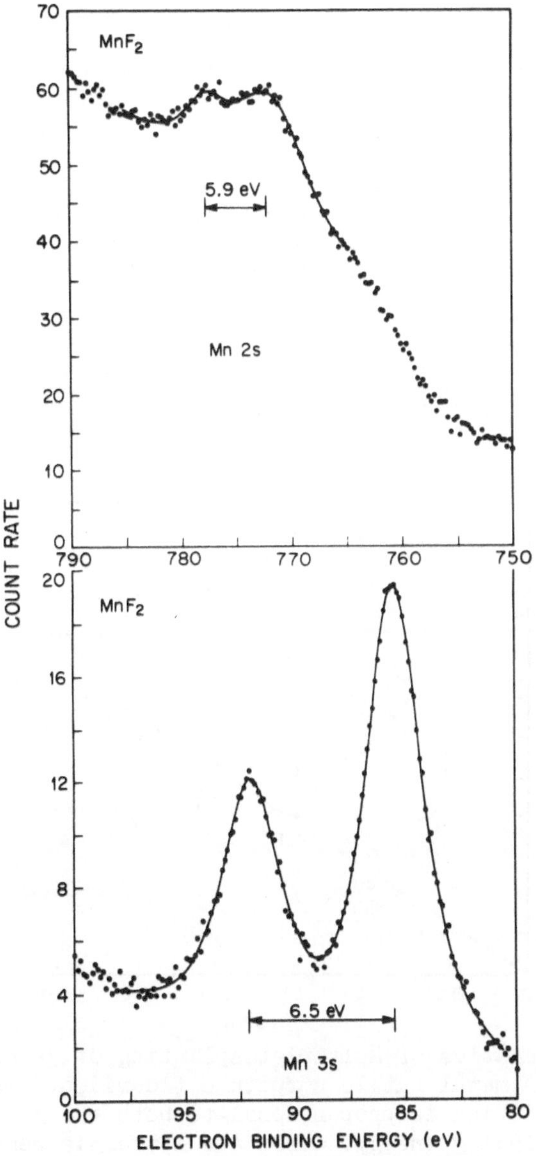

Fig. 3. Multiplet splitting of Mn2s and 3s-electrons in MnF_2. (from ref. 16)

The first goal is to show that s-electron multiplet splittings have systematic behavior with both spin of the outer shell and binding energy of the inner shell. This has been realized both for the 4s electrons of the rare earths (14), and for the 3s electrons of the transition metals, Fig. 4. In the rare earth trifluorides both the 4s and 5s splittings were found to be proportional to the spin of the 4f shell. The 4s splittings also explicitly exhibit the expected binding energy dependence. Hartree-Fock calculations show good agreement with the 5s splittings but yield values 75% larger than those measured for the 4s electrons. (15) This discrepancy may be due to the neglect of correlation and relaxation effects in the calculations. (15)

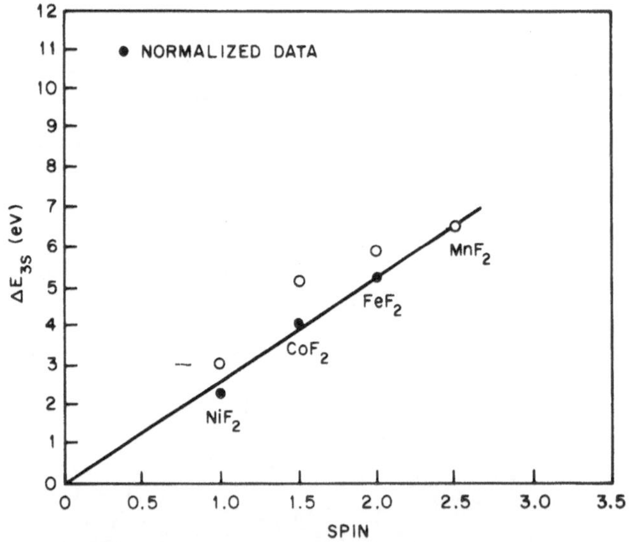

Fig. 4. Systematics of Multiplet splitting of 3s electrons in transition metal rutile structure fluorides. The measured splitting is proportional to both the 3d spin and the 3s binding energy. When the latter is removed the expected proportionality to spin is observed, see the solid points (from ref. 16)

In MnF_2 the 3s splitting is also greatly overestimated by free-ion Hartree-Fock calculations. (13) In that case it was shown that a cluster calculation for an octahedral MnF_6^{4-} group does yield good agreement with experiment. More recently the splitting of the 2s and 3s electrons of a number of 3d-group transition metal rutile structure fluorides has been obtained for a more detailed comparison with Hartree-Fock calculations. (16) The measured 2s splittings agree with the calculated values which differ little for free ion or cluster. The 3s splittings exhibit the expected spin and binding energy dependence. A major problem arises, however, when one tries to account for the hyperfine field in terms of the cluster calculation. The reduced 3s splitting results in a reduced 3s contribution to the contact interaction which makes the net calculated field large and negative and destroys the good agreement between theory and experiment.

There are two effects which may contribute to a resolution of this problem (1) one can invoke 4s electron covalency and let the strongly polarized 4s electron compensate for the loss of 3s electron spin density or (2) one can question the cluster calculation and attribute the discrepancy in multiplet splitting to the neglect of relaxation and correlation effects in the Hartree-Fock calculations. (15)

2. Isomer and Chemical Shifts

The Mössbauer effect isomer shift (17) provides a direct measure of the electronic charge density at the nucleus. The binding energy, E_b, of the atomic electrons is sensitive to the screening of the nuclear potential by other electrons, especially those which have charge density close to the nucleus. As a consequence one might expect a correlation between isomer shifts, IS, and XPS binding energy shifts, and could even hope to map out charge densities in atoms. A more careful look at the factors which determine electron binding energies will show, however, that this neglects the major source of XPS shifts, which arise from electrostatic effects of the outer electrons and the surrounding crystal.

For concreteness consider the electron binding energies in an ionic solid made up of M^+ and X^- ions. The E_b of an electron in an M^+ ion can be approximated by taking that in a neutral M atom and increasing it by the electrostatic potential produced inside the atom by the removal of a valence electron. The binding energies of the ions in the solid relative to the vacuum level are then obtained in the simplest, but quite adequate model by shifting the M^+ levels up and the X^- levels down by the Madelung energy.(18) In this approximation all levels of an ion are shifted by the same amount, a prediction which has been reasonably well verified by XPS. The Madelung shift in general cancels out a large fraction of neutral atom to free ion shift so that binding energies shift much less with changes in valence than one might have supposed. To introduce covalency into this model we could imagine that some s-like electron charge is transferred back to the M^+ ion. This, however, will change both the Madelung energy and the binding energies of the electrons, i.e. the effects of the crystal enter in an important way and in practice cancel most of the shift obtained from the single ion model.(19, 20) Concisely stated the dominant effect on core-electron binding energies may be thought of as arising from the change in the electrostatic potential produced by removing an outer electron from a spherical shell with suitable charge radius. Charge neutrality requires that this electron is not removed to infinity but appears somewhere else in the crystal. The resultant effect on the Madelung energy then cancels part of the shift calculated for the isolated ion.

In molecular electron spectroscopy excellent correlations between binding energies and covalency parameters has been obtained. (2) In solids this has proved to be more difficult, largely because of the charging effects described below. In general shifts toward greater E_b have been found with increasing oxidation state for metal ions. Measured shifts are of the order of 1 to 3 eV per oxidation state, much smaller than would be produced by the removal of an outer electron from a free ion. Unfortunately most of the existing data cannot be used as a basis for detailed analysis because charging effects have not been adequately characterized.

An example drawn from recent work on iron compounds will serve to illustrate these remarks. The positive ^{57}Fe isomer shift (increase in E_γ) of Fe^{2+} relative to Fe^{3+} is due to a decrease in the total s-electron charge density at the nucleus, ρ_o, resulting from shielding of the nuclear charge by the additional 3d electron. (17) Can we predict the effect on the E_b of a 3p electron in the same atom? Naively we might expect that decreased s-electron charge density will decrease the shielding of the nuclear charge resulting in increased E_b. This,however, is counter to director observation, (21) Fig. 5, which shows increasing E_b with increasing oxidation state, a quite general result. One might imagine that the observed effect is due to increased shielding of the nuclear charge by the added 3d electron. The additional observation that 2p and 3p E_b's shift by the same amount rules out 3d-electron shielding effects. The real mechanism is the electrostatic one due to the 3d electrons and surrounding crystal.

Comparisons of IS and E_b shifts have also been made for ^{197}Au and ^{119}Sn. In the case of Au in alloys and intermetallic compounds (22) increasing ρ_o was accompanied by increasing 4f E_b, clearly indicating that the increase in s-electron charge density is accompanied by a reduction in occupancy of some other level, here the 5d band. Detailed calculations showed a ratio of d-electron depletion to conduction electron gain of 0.6 ± 0.2 in AuAg alloys.

In a set of tin compounds with the same stereochemistry and oxidation number a linear relationship was found between the ^{119}Sn IS and the 4d E_b; increasing ρ_o corresponding to decreasing E_b. (23) This is consistent with the picture outlined above in that the increase in 5s electron occupancy with increasing covalency directly explains both the change in IS and E_b.

The fact that binding energy shifts are generally approximately the same for all the electronic levels of an atom is a clear indication that these shifts originate in electrostatic effects of the valence electrons and surrounding crystal. Small differences between core-electron shifts, in principle, do contain information about the distribution of electronic charge within the atoms. It appears at present that such differential shifts are more effectively obtained from shifts in the X-ray emission line involving two such levels. Observed chemical shifts of X-ray energies are orders of magnitude smaller than core-electron binding energy shifts, but have been measured quite successfully. (24-26) In summary, XPS sees the direct electrostatic effects of changes in outer shell

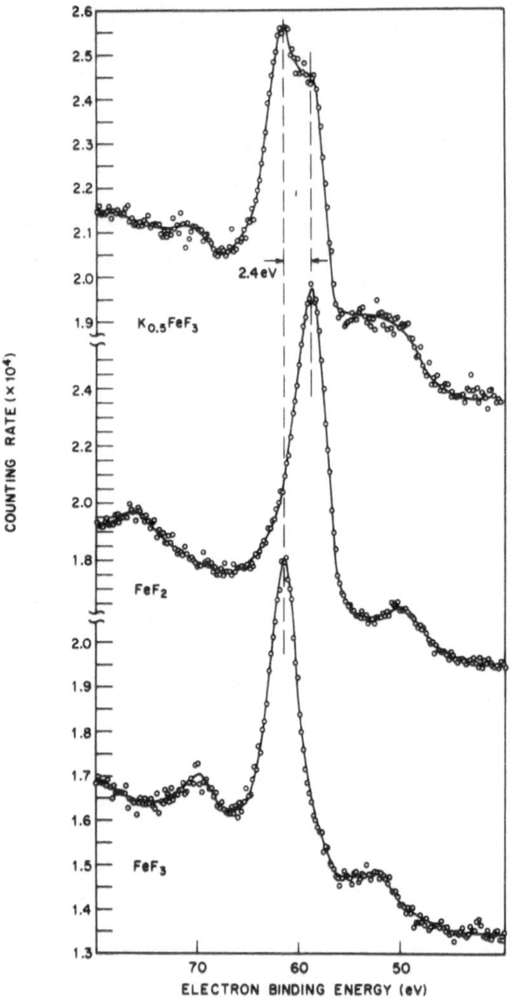

Fig. 5 Shift of XPS binding energy with change in valence in $K_{1/2}FeF_3$ which has the Na_xWO_3 structure. Since both the Fe^{2+} and Fe^{3+} coexist in the same crystal, uncertainties due to charging effects are avoided. Furthermore some ions of both valence occupy at least one of the two distinct lattice sites, but no splitting attributable to Madelung energy differences are observed. (from ref. 21)

occupancy; MS is sensitive to such changes only through the perturbation of the s-electron density at the nucleus; X-ray emission line shifts are sensitive to changes in charge distribution within the atom.

Some Cautionary Comments

Many experimental problems are encountered in XPS because it is basically a surface sensitive technique. In practice this means that few materials can be examined after exposure to room air without careful surface preparation in high vacuum. Only noble metals, and some fluorides, chlorides, and oxides appear to give satisfactory results without further treatment, but surface contamination is readily detected even on gold after a brief exposure to room air. Metals can be cleaned by ion bombardment, other materials may be cleaved or evaporated in vacuum, or cleaned by a variety of other techniques. (27)

A purely technical problem that has not been adequately resolved is that of "sample charging", encountered when applying XPS to insulators. Meaningful binding energy measurements can only be made if the sample and the spectrometer are at the same electrostatic potential. The surface of a dielectric sample is not readily grounded. Currently most spectrometers rely on low-energy electrons in the sample chamber to keep the surface near the potential of the surrounding sample chamber. Charging effects of a few volts are nevertheless commonly encountered. To deal with this problem it has been suggested that a reference substance be deposited on the surface of the sample, e.g. Au (28) or C. While this may provide a guide to the magnitude of the charging effects it must not be assumed that the Fermi level of the gold is coincident with that of the insulator.

Summary

X-ray photoelectron spectroscopy makes core and valence electron properties of solids accessible to direct experimental observation. Multiplet splittings of core s-electron shells relate directly to the core-polarization hyperfine fields seen in Mössbauer effect. Chemical shifts in XPS relate to isomer shifts only in the sense that both are traceable to valence shell occupancy. Differences between core-level shifts of two inner electrons are directly related to the charge density distribution in the atom, but are difficult to measure by XPS.

References

1. K. Siegbahn, et al., "ESCA, Atomic Molecular and Solid State Structure Studied by Means of Electron Spectroscopy" (Almqvist and Wiksells, Uppsala, 1967).
2. K. Siegbahn, et al., "ESCA Applied to Free Molecules", (North Holland, Amsterdam and London, 1970).
3. R. L. Mössbauer, Z. Physik 151, 124 (1958); Naturwissenschaften 45, 538 (1958), Z. Naturforsch. 14a, 211 (1959).
4. K. Siegbahn, et al., Science 176, 245 (1972).
5. R. P. Godwin, "Synchrotron Radiation as a Light Source" in Springer Tracts in Modern Physics 51, 2 (1969).
6. E. M. Purcell, Phys. Rev. 54, 818 (1938).
7. J. C. Helmer and N. H. Weichert, Appl. Phys. Letters 13, 266 (1968).
8. H. Kanter, Phys. Rev. B 1, 2357 (1970).
9. Y. Baer, et al., Solid State Communications 8, 1479 (1970).
10. R. G. Steinhardt, J. Hudis and M. L. Perlman, Phys. Rev. B 5, 1016 (1972).
11. W. N. Delgass, T. R. Hughes and C. S. Fadley, Catalysis Reviews 4, 179 (1970).
12. See for example: R. E. Watson and A. J. Freeman, Phys. Rev. 123, 2027 (1961).
13. C. S. Fadley, et al., Phys. Rev. Letters 23, 1397 (1969), C. S. Fadley and D. A. Shirley, Phys. Rev. A 2, 1109 (1970).
14. R. L. Cohen, G. K. Wertheim, A. Rosencwaig and H. J. Guggenheim, Phys. Rev. B 5, 1037 (1972).
15. J. F. Herbst, D. N. Lowy and R. E. Watson (to be published).
16. S. Hüfner and G. K. Wertheim (to be published).
17. L. R. Walker, G. K. Wertheim and V. Jaccarino, Phys. Rev. Letters 6, 98 (1961).
18. See for example, F. Seitz, "The Modern Theory of Solids" (McGraw-Hill, New York and London, 1940) Chapter II.
19. See ref. 1, Chapter V for a more detailed discussion of this problem.
20. C. S. Fadley, S. B. M. Hagström, M. P. Klein and D. A. Shirley, J. Chem. Phys. 48, 3779 (1968).

21. D. N. E. Buchanan, et al., Solid State Communications 9, 583 (1971).
22. R. E. Watson, J. Hudis and M. L. Perlman, Phys. Rev. B 4, 4139 (1971).
23. M. Barber and P. Swift, Chem. Comm. 1338 (1970).
24. O. I. Sumbaev, Soviet Physics, JETP 30, 927 (1970) and references cited therein.
25. B. K. Agarwal and L. P. Verma, J. Phys. C 1, 208 (1968); ibid. 2, 104 (1969); ibid. 3, 537 (1970).
26. B. G. Gokhale, R. B. Chesler and F. Boehm, Phys. Rev. Letters 18, 957 (1967).
27. M. Kaminsky "Atomic and Ionic Impact Phenomena on Metal Surfaces" (Academic Press, New York, 1965).
28. D. J. Hnatowich, et al., J. Appl. Phys. 42, 4883 (1971).

CONSEQUENCES OF NUCLEAR TRANSFORMATIONS IN CHEMICAL COMPOUNDS STUDIED BY THE MÖSSBAUER METHOD

Ursel Zahn, W. Potzel and F.E. Wagner
Physik-Department
Technische Universität München
D-8046 Garching, Germany

Chemical consequences of β^--decay, electron capture and (n,γ)-reactions in compounds of 4d and 5d transition elements have been studied at 4.2 K and 1.8 K by Mössbauer spectroscopy. Emission spectra of a number of oxo and chloro complexes are presented. Various decay products could be identified by their isomer shifts. The results indicate that the nature of the products formed depends on the chemical properties of the host matrix rather than on the preceeding nuclear process.

I. INTRODUCTION

The recoil chemical behaviour of nucleogenic atoms has been studied extensively by various physical and chemical techniques. Possible reactions of recoiling atoms depend strongly on the environment in which these atoms originate. Only in vacuum can the primary changes in the electronic structure of decaying atoms be preserved long enough to be studied by methods like radioactive recoil spectrometry (1,2). Quite generally, about 80% of the β^- transitions take place without electron loss, leading to singly charged positive ions. In the remainder of the atoms electron shake-off takes place due to the sudden non-adiabatic change of the nuclear charge. In this way multiply-ionized recoil atoms are formed. Multiple ionization also results from electron capture (EC) and internal conversion

(IC) processes, which remove inner shell electrons and give rise to vacancy cascades. The charge distribution observed after the isomeric transition (IT) of ^{131m}Xe (2), for instance, extends to +13 and has its maximum at +8. If the atoms in which the vacancy cascade occurs is one of the constituents of a molecule in a gas, extensive decomposition of this molecule takes place. Gaseous CH_3I, for instance, explodes due to Coulomb repulsion if vacancies are created in iodine by X-rays (3).

The picture becomes exceedingly more complex and less understood, if the nuclear transformations are taking place in condensed phases. First of all, ion-electron recombination reactions in liquids and solids take place on an extremely fast time scale of 10^{-14} s or less. Secondly, even if no instantaneous recombination occurs, molecular fragments left behind in the solid may decisively influence the reaction paths of the nucleogenic atoms during successive annealing steps and during analytical separation procedures. Analytical methods for the identification of the recoil products, such as dissolution, volatilization or chromatography, always necessitate a destruction of the immediate environment of the hot atom and at best allow indirect conclusions as to its initial state (4-6).

Therefore methods like perturbed angular correlations (PAC) and Mössbauer emission spectroscopy, which use γ-rays emitted after the nuclear event as a probe, were applied with great anticipation in recoil-chemical studies. By these techniques the products of nuclear transformations can be studied "in situ" without any interference from chemical analysis. With both methods one obtains information on the state of the nucleogenic atom in a time-interval between roughly 10^{-9} s and 10^{-6} s after the decay feeding the Mössbauer level or the γ-γ- cascade. The primary information obtained by both Mössbauer spectroscopy and PAC is on the hyperfine interactions of the nucleogenic atoms. A severe limitation of both methods is their limited resolution, which may render the interpretation of the data difficult or even impossible, particularly in cases where several different species are formed.

Here, Mössbauer spectroscopy has the advantage of measuring the isomer shift (IS) in addition to electric quadrupole and magnetic dipole interactions. From the IS the charge states of the nucleogenic atoms can be deduced in many cases. This knowledge may then be helpful in attempts to interpret any electric quadrupole and magnetic dipole interactions observed. Still, conclusions about the geometry and state of the environment of the recoil products should only be drawn with great care. Amongst the drawbacks of the Mössbauer technique is its limitation to low temperatures in studies with all but those transitions which have sufficiently low energy to ensure a reasonably high recoilless fraction even at room temperature or above. Entities which, due to severe rupture of their chemical bonds, have suffered a drastic decrease of the recoilless fraction, may escape observation completely. In comparisons of results from Mössbauer experiments with results obtained by chemical methods, one should also keep in mind, that the effects of decay paths which do not populate the Mössbauer level also escape observation by the Mössbauer method.

Nevertheless, Mössbauer emission spectroscopy has provided numerous results in recoil chemistry, mainly with respect to the possibility of an identification of the electronic structure of the recoil products. Preferentially the consequences of the EC decay of ^{57}Co in various matrices have been investigated. The aftereffects of a number of other transitions, like the IT of ^{119m}Sn and ^{129m}Te and the β^--decay of ^{129}I and ^{129}Te have also been investigated. The information obtained in these studies has been reviewed in several recent comprehensive articles (7-10). Sofar, however, little Mössbauer work on the recoil chemistry of 4d and 5d elements has become known.

In the present paper the results of recent studies of chemical effects of nuclear transformations in covalent compounds of such elements will be reported and the results will be discussed.

II. THE STUDIED TRANSITIONS AND THEIR PROPERTIES

The Mössbauer transitions in ^{99}Ru (89 keV), ^{193}Ir (73 keV) and ^{197}Au (77 keV) were used in our studies of the emission spectra of sources of ^{99}Rh, ^{193}Os, ^{197}Pt, and ^{197}Hg incorporated in various compounds. The high γ-ray energies made cooling to liquid He temperature imperative in all cases. The ^{99}Rh source activity was produced by the ^{99}Ru (d,2n)^{99}Rh reaction. All other activities were obtained by neutron irradiation. The Mössbauer level in ^{99}Ru is populated by EC and that in ^{193}Ir by β^--decay. The 77 keV level in ^{197}Au can be populated both by β^--decay from ^{197}Pt and by EC from ^{197}Hg. In all of these cases the IS by far exceed the experimental line width and is sufficiently well understood to be used in the assignment of oxidation states to nucleogenic species.

The IS of compounds of Ru, Os, Ir, and Au (11-15) reveal characteristic similarities, which can be understood in terms of shielding effects of the varying number of valence shell d electrons. Moreover, the increase of the electron density at the nuclei of the transition elements in the presence of backbonding ligands is reflected in their IS (12). The right sides of the columns in Fig.1 show IS for compounds of Ru, Os, Ir, and Au studied as absorbers. In these bar-diagrams the IS are referred to the respective metals and normalized in such a way that isoelectronic compounds like e.g. OsO_4 and RuO_4 or $K_4Os(CN)_6$ and $K_3Ir(CN)_6$ lie on horizontal lines. The s-electron densities at the metal nuclei are increasing from the bottom to the top of Fig.1. The respective d-electron configurations are indicated on the righthand side. The knowledge of the IS in a great number of stable compounds yields the basis for an interpretation of emission spectra of various radioactive labelled compounds. The left sides of the columns in Fig.1 contain the IS of tentatively formulated recoil species, and will be discussed in the following. It should be noted that the sign convention for all shifts given in Fig.1 and in the remainder of this paper is such, that shifts measured in sources are given with the sign actually found in the Mössbauer spectra, whereas for shifts measured in absorbers the sign is reversed in order to facilitate a comparison with the data obtained from the source experiments.

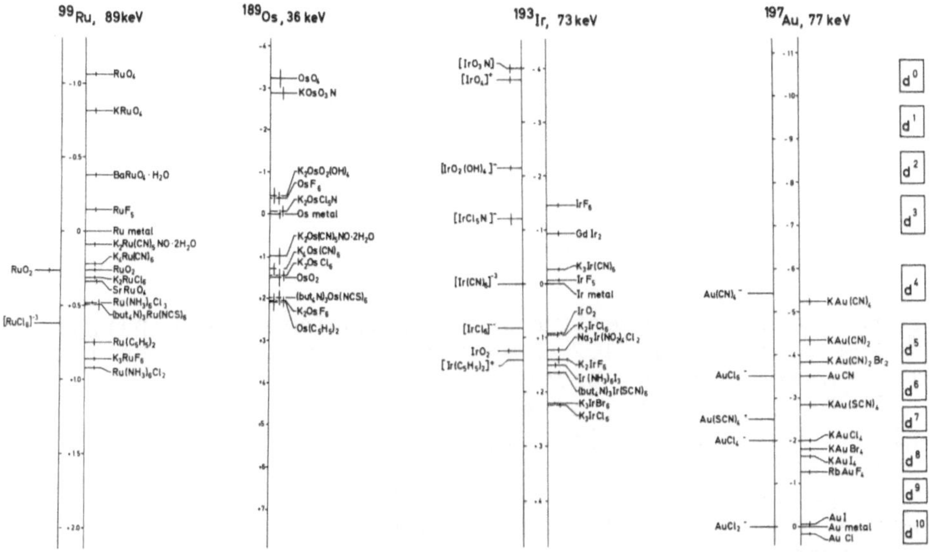

Fig.1: Summary of isomer shift results for Ru, Os, Ir, and Au, plotted in units of mm/s on vertical bar diagrams. Data for compounds studied as absorbers are given on the right sides of the columns. The results obtained from the emission spectra are given on the left sides. All shifts refer to the respective metals. They are normalized in such a way that isoelectronic compounds lie on horizontal lines and electron densities increase from the bottom to the top of the figure.

III. EMISSION SPECTRA OF COMPOUNDS OF OS(VIII) AND OS(VI)

Emission spectra of sources of OsO_4 (16), $KOsO_3N$, K_2OsCl_5N and $K_2[OsO_2(OH)_4]$ (16) are shown in Fig.2. The OsO_4 source was prepared by neutron irradiation of OsO_4 and subsequent purification by

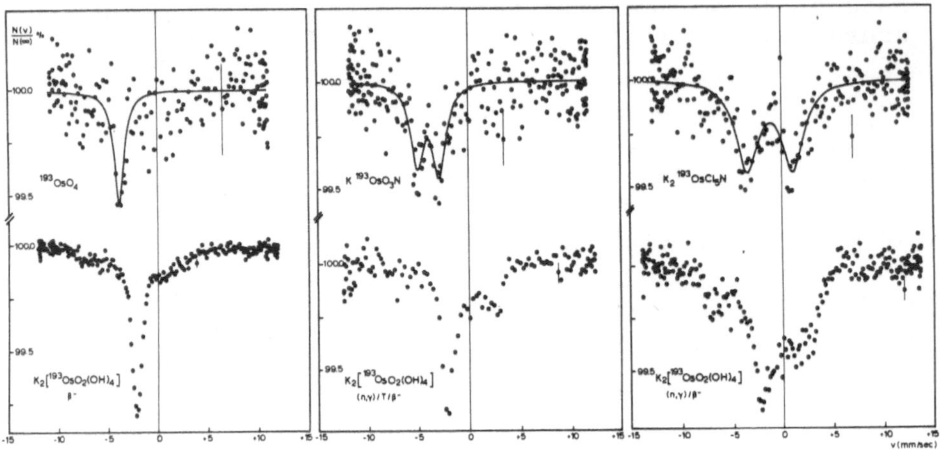

Fig.2: Emission spectra of Os compounds containing ^{193}Os measured at 4.2 K against an Ir metal absorber. The spectra in the bottom row are that of $K_2[OsO_2(OH)_4]$ prepared after the neutron irradiation from Os metal (left), and those of $K_2[OsO_2(OH)_4]$ prepared before the irradiation (middle and right). The spectrum in the middle was taken with a source which, additionally, was annealed at 150° C after the irradiation.

vacuum sublimation. The sources of $KOsO_3N$, K_2OsCl_5N and $K_2[OsO_2(OH)_4]$ were prepared from neutron - irradiated osmium metal, which was oxidized and destilled into the respective reagents (17). Fig.2 also shows two emission patterns of $K_2[OsO_2(OH)_4]$ sources which were irradiated after the preparation of the compound and have suffered the (n,γ)-process and a different annealing history.

The main primary products of the β^--decay of ^{193}Os are expected to be Ir species which are isoelectronic to the Os compound studied. OsO_4 and $KOsO_3N$ contain Os in the formal oxidation state +8 with a $5d^0$ electron configuration. In K_2OsCl_5N and $K_2[OsO_2(OH)_4]$ the Os is hexavalent with a $5d^2$ configuration. The IS of such Ir entities with $5d^0$ and

$5d^2$ electron configurations cannot be observed in absorption Mössbauer spectroscopy, since the highest oxidation state of Ir that could be prepared to-date is +6 with a $5d^3$ configuration. The IS expected for higher valence states of Ir can, however, be estimated by extrapolation from the shifts for lower oxidation states of Ir and by comparison with the shifts for isoelectronic Os compounds (Fig.1).

The prominent feature in the spectra of OsO_4 and $K_2[OsO_2(OH)_4]$ is a single emission line. The spectra of $KOsO_3N$ and K_2OsCl_5N exhibit two lines with nearly equal intensities. These are assumed to be quadrupole doublets. The IS observed in these spectra (Fig.1) show, that the recoil products of OsO_4 and $KOsO_3N$ are Ir^{+9} ($5d^0$) species, whereas the β^--decay in $K_2[OsO_2(OH)_4]$ leads to Ir^{+7} ($5d^2$). The assignment of a $5d^2$ configuration to the decay product of K_2OsCl_5N is somewhat ambiguous, since the IS could also be reconciled with a $5d^3$ configuration. The assignment of $5d^0$ and $5d^2$ configurations to the main decay products of OsO_4, $KOsO_3N$, K_2OsCl_5N, and $K_2[OsO_2(OH)_4]$ is, however, supported by the absence of magnetic hyperfine splittings in the emission patterns. Species with odd electron configurations, i.e. with Kramers-degenerate groundstates, should exhibit paramagnetic hyperfine patterns in the diamagnetic host lattices of the studied sources. Species exhibiting magnetic hyperfine interactions may, however, contribute to the broad unresolved background which becomes noticeable with improved data statistics, e.g. in the emission spectrum of $K_s[OsO_2(OH)_4]$ (Fig.2).

The tetrahedral symmetry of the OsO_4 molecule seems to be preserved in the case of the $(IrO_4)^+$ entity. The quadrupole splittings observed in the decay products of $KOsO_3N$ and K_2OsCl_5N sources are reminiscent of those found in these compounds used as absorbers for Mössbauer measurements with the 69.6 keV γ-rays of ^{189}Os (18). The single line in the emission spectrum of the $K_2[OsO_2(OH)_4]$ source, however, is surprising since $K_2[OsO_2(OH)_4]$ has a distorted octahedral symmetry and its absorption Mössbauer spectrum (18) indeed exhibits a quadrupole splitting. The single-line pattern may be due to an accidental cancellation of the contributions of the different 5d orbitals to the electric field gradient, but it could also be caused by a local structural change like the formation of a tetrahedral $(IrO_4)^-$ entity.

The (n,γ)-reactions in neutron-activated $K_2[OsO_2(OH)_4]$ lead to a poorly resolved spectrum (Fig.2) with still an indication of the single line predominant in the $K_2[OsO_2(OH)_4]$ source which has undergone β^--decay only. This peak becomes more pronounced again if the source is annealed at 150° C after the neutron irradiation.Which indicates that at that temperature $[OsO_2(OH)_4]^{2-}$ molecules are formed from the fragments produced during the (n,γ)-reaction. The broad unresolved patterns observed with the neutron irradiated $K_2[OsO_2(OH)_4]$ sources may partly be due to paramagnetic hyperfine structure, e.g. of Ir^{+6} species, but more experiments and better statistics will be needed to clarify this point.

IV. EMISSION SPECTRA OF RUTILE-TYPE DIOXIDES

Sources of ^{99}Rh in RuO_2, ^{193}Os in OsO_2 and ^{197}Pt in PtO_2 were produced by (d,2n) and (n,γ) reactions, respectively. Their emission spectra are shown in Fig.3. Results obtained by least squares fits of quadrupole patterns to the data are given below together with values observed in absorption experiments.

Source	Absorber	IS (mm/s)	$eQV_{zz}/2$ (mm/s)
$^{99}Rh/RuO_2$	Ru metal	$+0.28 \pm 0.01$	0.47 ± 0.02
$^{99}Rh/Ru$	RuO_2	$+0.26 \mp 0.01^+$	0.50 ∓ 0.01
$^{193}Os/OsO_2$	Os metal	$+1.25 \pm 0.02$	2.41 ± 0.02
$^{193}Os/Os$	OsO_2	$+0.94 \mp 0.02^+$	2.47 ∓ 0.02

($^+$ IS values with respect to the metals and with the sign reversed to facilitate the comparison with the source experiments).

RuO_2, OsO_2 and IrO_2 have the tetragonal rutile lattice and are conductors because of their incompletely

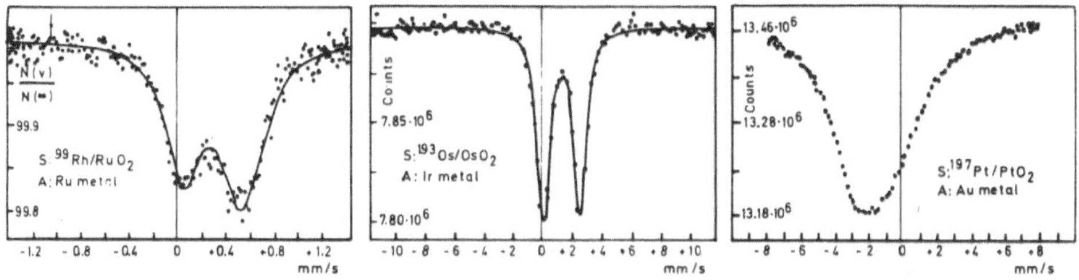

Fig.3: Emission spectra of sources of ^{99}Rh in RuO_2, ^{193}Os in OsO_2 and ^{197}Pt in PtO_2 measured at 4.2 K against Ru, Ir, and Au metal absorbers, respectively.

filled 5d (t_{2g}) bands (19). PtO_2 is hexagonal and an insulator due to its $5d^6$ electron configuration. The emission spectra of ^{99}Rh in RuO_2 and ^{193}Os in OsO_2 show a pure quadrupole pattern. This indicates that any defects produced by the nuclear processes in the MeO_2 lattice have healed out by the time of the emission of the Mössbauer γ-rays and that neither the preceeding nuclear reactions nor the following EC or β^--transitions result in stable aliovalent charge states of daughter atoms. Any such effects could, indeed, hardly be reconciled with the metallic conductivity of these compounds.

Expectedly, the IS for the source of RuO_2 corresponds exactly to that observed in a RuO_2 absorber. The IS for ^{193}Ir originated in OsO_2, however, shows that the electron density at Ir(IV) in OsO_2 is smaller than the electron density in IrO_2 by about 1/3 of the difference between Ir(IV) and Ir(III). This indicates that charge screening in the OsO_2 lattice leads to an accumulation of d charge on Ir. In neutron-irradiated PtO_2 an unresolved ^{197}Au emission spectrum is observed (Fig.3) that shows the presence of several ^{197}Au species. The IS indicates that the +3 charge state is formed preferentially, but the additional formation of other charge states cannot be ruled out. It would well be conceivable in the insulating PtO_2.

V. EMISSION SPECTRA OF METAL HEXAHALIDES

a. Sources of $^{193}OsCl_6^{2-}$

Emission spectra of ^{193}Os incorporated in hexachlorometallates of osmium(IV), platinum(IV), and iridium(IV) measured against an Ir metal absorber are shown in Figs. 4 and 5.

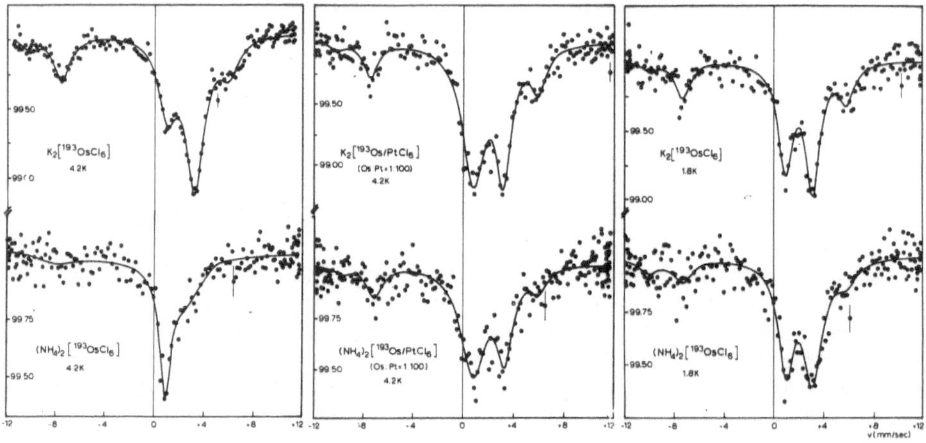

Fig. 4: Emission spectra of K_2OsCl_6 and $(NH_4)_2OsCl_6$ at 4.2 K and 1.8 K, and of K_2OsCl_6 diluted in K_2PtCl_6 (Os:Pt=1:100) at 4.2 K. The absorber always was Ir metal.

The sources were prepared by neutron activation of K_2OsCl_6 and $(NH_4)_2OsCl_6$. The irradiated samples were annealed for 2 h at 150° C and then, for further purification, dissolved in 0.1 n HCl, filtered, and reprecipitated by addition of ethanol. The dilution of $^{193}OsCl_6^{2-}$ in $PtCl_6^{2-}$ or $IrCl_6^{2-}$ (Os : Pt = 1:100; Os : Ir = 1 : 75) was achieved by adding the filtrate containing the $^{193}OsCl_6^{2-}$ solution to concentrated solutions of H_2PtCl_6 or H_2IrCl_6 in 0.1 n HCl and precipitating the salts with a small excess of NH_4Cl or KCl.

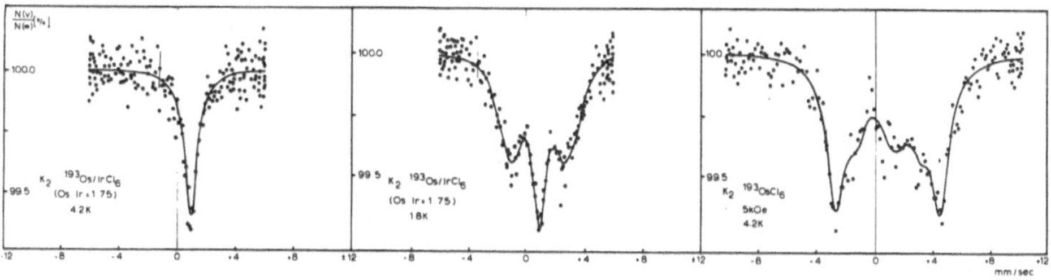

Fig.5: Emission spectra of K_2OsCl_6 diluted in K_2IrCl_6 (Os : Ir = 1 : 75) at 4.2 K and 1.8 K and of pure K_2OsCl_6 at 4.2 K in a longitudinal external magnetic field of 5 kOe. The absorber always was Ir metal.

The spectra of Figs.4 and 5 all arise essentially from Ir^{+4}, but sensitively reflect the magnetic properties of the host lattices. The hexahalides of Os, Ir and Pt have the cubic hexachloroplatinate (IV) structure with almost identical lattice parameters. K_2PtCl_6 and $(NH_4)_2PtCl_6$ have low-spin $5d^6$ electron configurations and are diamagnetic. K_2IrCl_6 and $(NH_4)_2IrCl_6$ ($5d^5$), having Kramers degenerate groundstates, are paramagnetic at high temperatures and become antiferromagnetic below 3.08 K and 2.16 K, respectively. K_2OsCl_6 and $(NH_4)_2OsCl_6$ ($5d^4$) have a singlet electronic groundstate (20) and exhibit Van Vleck paramagnetism.

The primary product of β^--transitions in hexachloroosmate(IV) will be the isoelectronic $(IrCl_6)^-$ ($5d^4$) species which should give rise to a single line with an IS close to that of IrF_5, i.e. near zero velocity (Fig.1). None of the spectra of the $^{193}OsCl_6^{2-}$ sources (Figs.4 and 5) exhibits such a line. All spectra can be interpreted as a superposition of a single line near +1 mm/s and a magnetic hyperfine pattern which also has an IS of about +1 mm/s, typical for Ir^{+4} (Fig.1). The $(IrCl_6)^-$ entities thus turn out to undergo rapid reduction to the chemically stable Ir^{+4} ($5d^5$) configuration.

For isolated $IrCl_6^{2-}$ impurities in the diamagnetic hexachloroplatinates the spin-lattice relaxation time at 4.2 K and below is known to be sufficiently small (21) for a paramagnetic hyperfine pattern to be seen in the Mössbauer spectra and the same should be the case in the hexachloro-osmate host lattice. In cubic symmetry the hyperfine splitting in the paramagnetic state can be described by the Hamiltonian $\mathcal{H}=A\cdot\vec{I}\cdot\vec{S}'$, where $\vec{I}$ is the nuclear and $\vec{S}'$ the effective electronic spin. In the case of low spin $^{193}Ir^{+4}$, where $I_g=3/2$, $I_{ex}=1/2$ and $S=1/2$, this interaction splits the ground and excited nuclear state into two sublevels each. This situation is largely analogous to low spin Fe^{+3} ($3d^5$) in cubic symmetry (22), except that for the mixed E2/M1 transition in ^{193}Ir none of the transitions between the sublevels is forbidden. One hence expects the paramagnetic hyperfine pattern to consist of four lines. The spectra of Fig.4 were least-squares fitted with such a pattern and an additional single line. With the ratio of the nuclear g-factors restrained to $g_{ex}/g_g=9.82$ (23) and free line intensities, values between -79 MHz and -81 MHz were obtained for A_g, in excellent agreement with the ENDOR value of $A_g=-(79.536\pm0.010)$ MHz (24) for the ^{193}Ir groundstate in $(NH_4)_2\underline{Ir}\underline{Pt}Cl_6$.

In Fig.5 a spectrum of a K_2OsCl_6 source at 4.2 K in a longitudinal external magnetic field of 5 kOe is shown. In this case the nuclear and electronic spins are decoupled (22) and one observes a hyperfine slitting which is practically identical with a normal Zeeman pattern in a longitudinal field. A normal Zeeman pattern with an additional unsplit line is also observed in the spectrum of K_2OsCl_6 in antiferromagnetic K_2IrCl_6 at 1.8 K. Above the Néel temperature, at 4.2 K, this source exhibits a single line at $+(0.93\pm0.03)$ mm/s, typical for Ir^{+4}. Here, no paramagnetic hyperfine splitting occurs due to the fast spin-spin relaxations. The Zeeman pattern observed with the $K_2Os/\underline{Ir}Cl_6$ source at 1.8 K corresponds to a hyperfine field of (348 ± 9) kOe. This is considerably smaller than the field of (395 ± 5)kOe (23) observed in a K_2IrCl_6 absorber. The IS found for the split patterns are all between +0.9 and 1.1 mm/s. Those for the unsplit line tend to be slightly larger , as can be seen from the slight asymmetries of the spectra with Zeeman-type splitting (Fig.5). Still, the IS rules out the possibility that the unsplit line is due to diamagnetic Ir^{+3}. No convincing explanation can be given at this time for the fact that some of the Ir^{+4} species do not take part in the magnetic phenomena. These species may be associated with lattice imperfections which would also be present without the β^--decay. The existence of unsplit lines in the Mössbauer spectra of hexachloroiridate absorbers below the Néel temperature (15) supports such a view. The relative intensities of the split and unsplit patterns might then be strongly dependent on such details of the source preparation as the speed of crystallization, and the observed differences of the intensity ratios of the split and unsplit part of the spectra would be largely fortuitous. It should be pointed out that the present data do not give evidence for a temperature dependence of these intensity ratios, since spectra for different temperatures were also taken with different sources. The 12% reduction of the hyperfine field in $K_2Os/\underline{Ir}Cl_6$ as compared to that in K_2IrCl_6 absorbers also cannot readily be explained. It may be due to the statistically distributed Os impurities in the $K_2Os/\underline{Ir}Cl_6$ lattice, which affect the exchange interaction. If this is the case, the field of 348 kOe should rather be regarded as an average over a distribution of hyperfine fields.

b. Sources of $^{197}PtCl_6^{2-}$ and $^{99}RhCl_6^{2-}$

The emission spectra reproduced in Fig.6 reflect the consequences of the β^--decay of ^{197}Pt and of the EC decay of ^{99}Rh in diamagnetic hexachloro matrices. The platinum compounds were synthesized from neutron activated platinum metal (17). To obtain $K_3{}^{99}RhCl_6$, ^{99}Rh from the ^{99}Ru (d,2n)^{99}Rh reaction had to be separated from the Ru target. Then, after addition of Rh metal carrier, K_3RhCl_6 was prepared by fusion with KCl in a stream of dry Cl_2(17).

The β^--decay in $(NH_4)_2PtCl_6$ leads to a broad and unresolved emission pattern, which has its emission maximum at a velocity where Au(V) ($5d^6$) would be expected by extrapolation from the positions of AuCl and $KAuCl_4$ (Fig.1). This extrapolation is supported by the recent observation of Au(V) in absorber experiments (25)

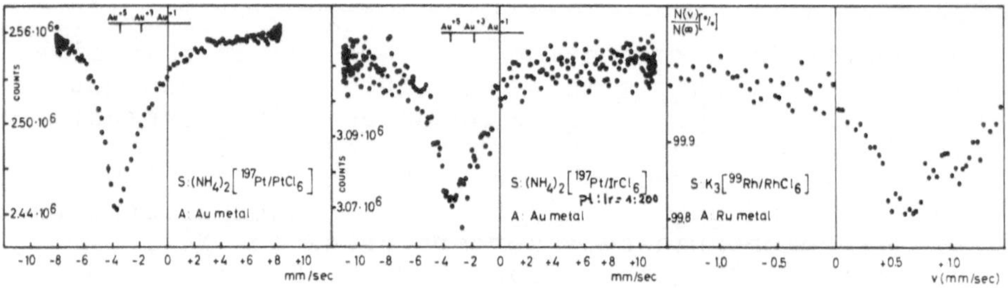

Fig.6: Emission spectra of $(NH_4)_2PtCl_6$, of $(NH_4)_2PtCl_6$ diluted in $(NH_4)_2IrCl_6$ (Pt:Ir = 1:200) and of K_3RhCl_6 measured at 4.2 K against Au and Ru metal absorbers, respectively.

with hexafluoro complexes of gold (26). One hence concludes that the isoelectronic $(AuCl_6)^-$ entity is sufficiently stable to dominate in the Mössbauer spectrum. The shoulder at higher velocities indicates the additional formation of Au(III) ($5d^8$) and Au(I) ($5d^{10}$) species. The incorporation of ^{197}Pt into the isostructural, but paramagnetic, $(NH_4)_2IrCl_6$ (Pt:Ir = 1:200) does not noticeably change the emission pattern (Fig.6). This indicates that no Au species with an odd number of electrons contribute to the unresolved spectrum, since these should exhibit a paramagnetic hyperfine splitting in the diamagnetic K_2PtCl_6, where relaxation times are long, but not in the paramagnetic K_2IrCl_6.

The emission spectrum of ^{99}Rh in K_3RhCl_6 is of relatively poor statistics and cannot be interpreted conclusively. It shows two emission peaks, a pattern which would be compatible both with the formation of Ru species in the +3 and the +2 charge state or with the formation of isolated centers of $RuCl_6^{-3}$ ($4d^5$), which could exhibit paramagnetic hyperfine interactions in the diamagnetic K_3RhCl_6 host lattice.

VI. EMISSION SPECTRA OF COMPOUNDS OF PT(II), HG(II) AND OS(II)

a. Sources of ^{197}Pt(II)- and ^{197}Hg(II)-Compounds

The emission spectra of the square planar

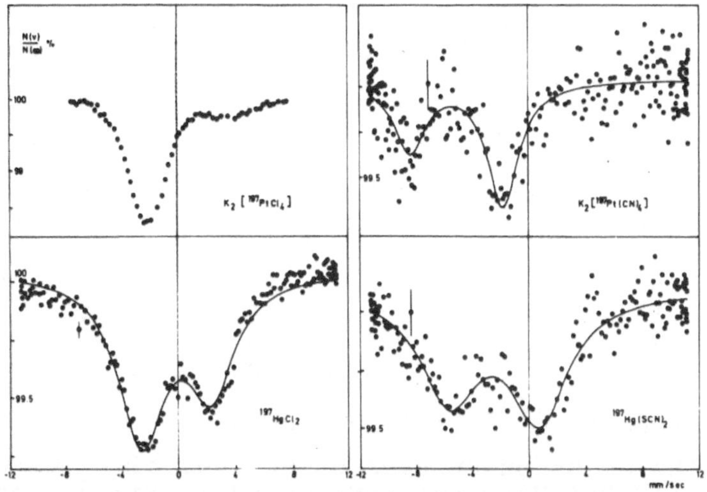

Fig.7: Emission spectra of divalent platinum and mercury compounds measured at 4.2 K against a Au absorber.

compounds K_2PtCl_4 and $K_2Pt(CN)_4$, which both have a $5d^8$ electron configuration, are shown in Fig.7. K_2PtCl_4 was prepared from neutron irradiated Pt metal by reduction of the intermediate product K_2PtCl_6 (17). The emission spectrum of K_2PtCl_4 exhibits a broadened emission line with the minimum at -2.0 mm/s, which indicates the predominant formation of Au^{+3} species, since the observed IS is very close (Fig.1) to the shift observed in a $KAuCl_4$ absorber. The $K_2Pt(CN)_4$ source, prepared by neutron irradiation of the compound and subsequent annealing, exhibits a quadrupole split pattern with a large negative IS of -5.4 mm/s. Both IS and quadrupole splitting agree well with the values observed for $KAu(CN)_4$ as an absorber. This shows that the β^--decay in $K_2Pt(CN)_4$ essentially leads to the formation of homologous $[Au(CN)_4]^-$ entities. The observed broadening of the emission lines may be due to the presence of ligand deficient or otherwise distorted entities, which might already have originated during the (n,γ)-reaction.

The $^{197}HgCl_2$ and $^{197}Hg(SCN)_2$ sources were made by neutron irradiation of the compound and subsequent recrystallization. In the emission spectrum of $^{197}HgCl_2$

a quadrupole pattern attributable to Au(I) ($5d^{10}$) on the basis of its IS of -(0.03±0.06) mm/s is observed. The spectrum of the $Hg(SCN)_2$ source consists of a pattern, whose IS of -(2.4±0.1) mm/s could be characteristic for both Au(I) and Au(III), if back-donation effects are taken into account. Together with the quadrupole splitting of 6.4±0.2 mm/s, however, this pattern can unambiguously (14) be assigned to Au(I). Neither in the $HgCl_2$ nor in the $Hg(SCN)_2$ source could higher charge states of gold be observed within the limits imposed by data statistics, although the EC decay populating the Mössbauer level must primarily have created highly charged ions.

b. Sources of ^{193}Os(II)-Compounds

The emission spectra of sources of $^{193}Os(C_5H_5)_2$ and $K_4[^{193}Os(CN)_6]$ are reproduced in Fig.8. Both compounds were neutron-activated after preparation. $Os(C_5H_5)_2$ was then purified by vacuum sublimation, which ensures a complete separation of undamaged osmocene from any recoil fragments. $K_4[^{193}Os(CN)_6]$ was merely annealed for 2 h at 120° C.

The β^--decay product of ^{193}Os bound in osmocene can be identified as the isoelectronic iridicinium-cation, $[Ir(C_5H_5)_2]^+$ by its IS and quadrupole splitting (16). The β^--decay of ^{193}Os in $K_4[Os(CN)_6]$ seems to lead essentially to the formation of isoelectronic Ir^{+3} entities (Fig.1). The observed line broadening

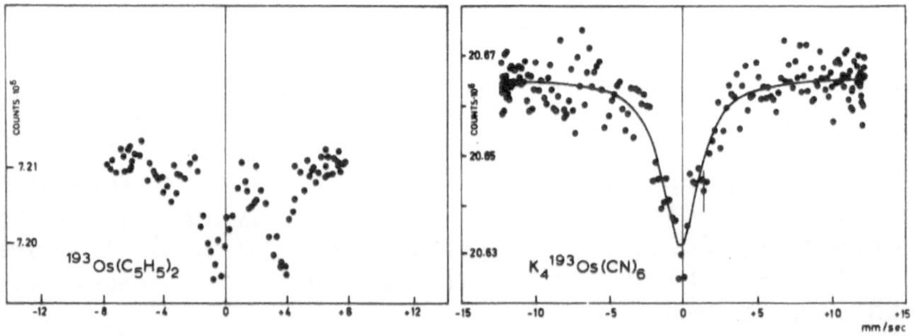

Fig.8: Emission spectra of sources of osmocene and $K_4Os(CN)_6$ measured at 4.2K against an Ir metal absorber.

may be due to the presence of distorted or ligand deficient entities which exhibit electric quadrupole splittings. It is conceivable that these entities are formed during the preceeding (n,γ)- reaction rather than by the β^--decay.

VII. CONCLUSIONS

The experiments and results reported in the preceeding sections are summarized in Tab.1. Although more data would be desirable and although the assignments of charge states may, in a few cases, still be ambiguous, some conclusions can be drawn as to the phenomenology and the mechanisms of product formation after nuclear transformations in compounds of the heavier transition elements. It emerges that in most cases the daughter product formed is the one most stable in the lattice environment of the parent compound, irrespective of the nature of the decay. In the case of the $HgCl_2$ and $Hg(SCN)_2$ sources this implies that irreversible disruption of the environment in these solids is a rare process even after the occurence of a vacancy cascade. Similar observations have been made in other cases, e.g. after the IT of ^{129m}Te in $(NH_4)_2TeCl_6$ (27).

During β^--decays the electronic structure of the parent atoms remains unchanged in roughly 80% of the cases. This fact has been used previously to produce species like $XeCl_4$, which had not yet been obtained by preparative methods when it was first observed by the Mössbauer effect after the β^--decay of ^{129}I in $KICl_4$ (28).

The present results confirm the idea that, if the daughter element forms stable complexes or molecules that are isostructural and isoelectronic with the parent compound, these will generally be the observed nucleogenic entity. There are several examples for this amongst the experiments reported here (Tab.1). In the square-planar K_2PtCl_4 and $K_2Pt(CN)_4$ complexes the isoelectronic Au species known from $KAuCl_4$ and $KAu(CN)_4$ are formed. In $K_4Os(CN)_6$ sources isoelectronic ($Ir(CN)_6^{3-}$), also known in iridium compounds like $K_3Ir(CN)_6$, is observed. ^{193}Os in osmocene decays to the iridicinium ion (16).

Nuclear Transformation	Source			Main Products		
	host-compound	FOS	d^n_{host}	FOS	d^n_{Prod}	remarks
$^{193}Os(\beta^-)\,^{193}Ir$	OsO_4	Os + 8	$5d^0$	Ir + 9	$5d^0$	(a)
	$KOsO_3N$	Os + 8	$5d^0$	Ir + 9	$5d^0$	-
	K_2OsCl_5N	Os + 6	$5d^2$	Ir + 7	$5d^2$	-
	$K_2[OsO_2(OH)_4]$	Os + 6	$5d^2$	Ir + 7	$5d^2$	(b)
	K_2OsCl_6	Os + 4	$5d^4$	Ir + 4	$5d^5$	(c)
	$K_2Os/\underline{Pt}Cl_6$	Os + 4	$5d^6$	Ir + 4	$5d^5$	(c)
	$K_2Os/\underline{Ir}Cl_6$	Os + 4	$5d^5$	Ir + 4	$5d^5$	(d)
	$(NH_4)_2OsCl_6$	Os + 4	$5d^4$	Ir + 4	$5d^5$	(c)
	$(NH_4)_2Os/\underline{Pt}Cl_6$	Os + 4	$5d^6$	Ir + 4	$5d^5$	(c)
	OsO_2	Os + 4	$5d^4$	Ir + 4	$5d^5$	(e)*
	$Os(C_5H_5)_2$	Os + 2	$5d^6$	Ir + 3	$5d^6$	-
	$K_4[Os(CN)_6]$	Os + 2	$5d^6$	Ir + 3	$5d^6$	- **
$^{99}Rh(EC)\,^{99}Ru$	K_3RhCl_6	Rh + 3	$4d^6$	Ru + 3	$4d^5$	(f)
	RuO_2	Ru + 4	$4d^5$	Ru + 4	$4d^5$	(e)*
$^{197}Pt(\beta^-)\,^{197}Au$	$(NH_4)_2PtCl_6$	Pt + 4	$5d^6$	Au + 5	$5d^6$	(g)
	$(NH_4)_2Pt/IrCl_6$	Pt + 4	$5d^6$	Au + 5	$5d^6$	(g)
	K_2PtCl_4	Pt + 2	$5d^8$	Au + 3	$5d^8$	(h)
	PtO_2	Pt + 4	$5d^6$	Au + 3	$5d^8$	(g)*
	$K_2[Pt(CN)_4]$	Pt + 2	$5d^8$	Au + 3	$5d^8$	**
$^{197}Hg(EC)\,^{197}Au$	$HgCl_2$	Hg + 2	$5d^{10}$	Au + 1	$5d^{10}$	-
	$Hg(SCN)_2$	Hg + 2	$5d^{10}$	Au + 1	$5d^{10}$	-

Table 1: Summary of experiments performed to study the chemical consequences of the decays populating the Mössbauer levels in ^{99}Ru, ^{193}Ir and ^{197}Au. The formal oxidation states (FOS) of the transition elements and the valence shell d electron configurations are given for the host compounds as well as for the predominant decay products.

(a) Broad background observed in addition to Ir^{+9} line.

(b) Decay product exhibits no quadrupole splitting, despite distortion of octahedral symmetry in the source compound.

(c) Superposition of a single line characteristic for Ir^{+4} and the paramagnetic hyperfine pattern of isolated $IrCl_6{}^{2-}$ complexes.

(d) Single line spectrum at 4.2 K, partly antiferromagnetic order at 1.8 K.

(e) Absence of aftereffects attributable to metallic conductivity.

(f) Ru^{+3} is the most probable product, the formation of some Ru^{+2} cannot be ruled out.

(g) Predominant product, but spectrum contains some contributions from lower oxidation states.

(h) Strongly broadened line.

* Sources prepared by preceeding nuclear reactions in the respective host lattices and no further processing.

** Sources prepared by direct neutron activation of the host compound and subsequent thermal annealing.

While these results are not unexpected, it is surprising that the EC decay of ^{197}Hg in $HgCl_2$ and $Hg(SCN)_2$ also leads to isoelectronic Au(I) species, since here highly charged Au must have been the immediate decay product. The formation of Au(I) reflects the similarity of the typical molecular structure of Hg(II) and Au(I) compounds. $HgCl_2$ forms crystals made up of linear Cl-Hg-Cl units. A linear X-Au-X structure is also typical for Au(I) in the halides (14) and in compounds like $KAu(CN)_2$. Hence the matrix structures of $HgCl_2$ and $Hg(SCN)_2$ will favour Au(I) species and prevent the stabilization of higher charge states of Au, which would require square-planar (Au(III)) or octahedral (Au(V)) environments.

In K_2PtCl_6 the nucleogenic Au does occupy octahedral lattice sites and, indeed, Au(V) has been observed in this case, although no stable $(AuCl_6)^-$ is known. Only very recently could Au(V) at all be prepared by chemical methods as the $(AuF_6)^-$ ion (26). The stability of Au(V) in K_2PtCl_6 can be explained by the non-availability of adequate electronic orbitals for the accommodation of additional electrons. For Au(V) ($5d^6$) in octahedral symmetry, the t_{2g} shell is filled. Any additional electrons would have to go into the e_g orbitals. This would be energetically unfavourable. In the square-planar Au(III) compounds the strong tetrahedral part of the ligand field splits the e_g orbitals into an upper one, b_{1g} ($d_{x^2-y^2}$), and a lower one, $a_{1g}(d_{z^2})$. The latter is sufficiently depressed to accommodate two additional electrons and thus favour a $3d^8$ configuration. In the K_2PtCl_6 lattice no such distortion seems to be possible, since Au(V) is stabilized.

The situation encountered after the β^--decay of ^{193}Os in K_2OsCl_6 and the lattices isostructural to it, is less readily explained. Although Ir(V) ($5d^4$) is known only in IrF_5 and as the $(IrF_6)^-$ anion, Ir(V) might have been expected to persist after the β^--decay of ^{193}Os. The results, however, show that the nucleogenic Ir undergoes a fast reduction to the chemically stable Ir(IV) ($5d^5$). This reduction may be facilitated by the presence of the electro-negative Cl^- ligands and the overlap of their p_π orbitals with the d_π wave functions of Ir.

It is the more surprising that Ir(VII) ($5d^2$), altogether unknown in chemically stable Ir compounds, persists in K_2OsCl_5N and $K_2[OsO_2(OH)_4]$ without having been reduced by the time of the emission of the Mössbauer gamma-ray.

Although the sort of argument made in the K_2PtCl_6 case is less convincing here, it is supported by the molecular-orbital scheme that has been suggested (29) for $K_2[OsO_2(OH)_4]$ and should also be applicable to K_2OsCl_5N. It has been assumed that the strong tetragonal part of the ligand field splits the t_{2g} orbitals into a doublet (e_g) and a singlet (b_{2g}), with the latter much lower than the former and occupied by the two d electrons of Os(VI). With the separation between e_g and b_{2g} large enough, this can explain the diamagnetism of $K_2Os(OH)_4$ and K_2OsCl_5N. Any additional electrons would have to be accommodated in the e_g orbitals. Since these are energetically unfavourable, this MO scheme could also explain the stability of Ir(VII) in these complexes.

Similar arguments may explain the stability of Ir(IX) in the tetrahedral coordination encountered in OsO_4 and $KOsO_3N$. The $5d^0$ configuration seems to be highly favoured in $(MO_4)^{n-}$ complexes of 5d elements. It is, indeed, the only configuration known in tetrahedral tetraoxo complexes of W, Re, and Os. The stabilization of Ir(IX) in the tetrahedral environment of OsO_4 and $KOsO_3N$ thus becomes conceivable.

One should, however, keep in mind that in addition to the species discussed sofar, others may be formed, albeit with smaller probabilities. Indications of such entities were observed in some of the emission spectra and probably would also become visible in at least some of the others with improved data statistics. These backgrounds may arise from bond rupture and autoradiolysis (30,31) after the occurence of vacancy cascades.

The experiments with $K_2[OsO_2(OH)_4]$ sources whose lattices have suffered a preceeding neutron irradiation show severe radiation damage, which begins to heal out only at elevated temperatures. The dioxides, particularly the electric conductors OsO_2 and RuO_2, do not show any radiation damage after the (n,γ) and (d,2n) reactions, respectively. This again reflects the importance of the matrix for the formation of different nucleogenic species.

Summing up, we find that in most of the cases reported in the present work, the chemical stability of the daughter products in the environment of the parent lattice has a paramount influence on the oxidation states formed after nuclear decays. More experiments are needed, however, before it will be possible to clearly assess the importance of effects like autoradiolysis or the influence of the lattice energy and size of the parent compound, all of which have been reported (30-33) to have some bearing on the consequences of nuclear transformations in solids.

(1) A.H.Snell, F.Pleasonton, and T.A.Carlson, in: "Chemical Effects of Nuclear Transformations", IAEA, Vienna 1961, Vol.1, p.146
(2) S.Wexler in M.Haissinsky, "Actions Chimiques et Biologiques des Radiations", Mason et Cie.,1965, p.107
(3) T.A.Carlson, and R.M.White, in: "Chemical Effects of Nuclear Transformations", IAEA, Vienna 1965, Vol.1, p.23
(4) A.P.Wolf, in:"Advances in Physical Organic Chemistry", V.Gold edt.,Academic Press, London,1964, Vol.2, p.201
(5) Contributions in "Nuclear Transformations in Solids" G.Harbottle, and A.G.Maddock, eds., North-Holland, Publishing Company, Amsterdam, in press
(6) "Chemical Effects of Nuclear Transformations", IAEA, Vol.I and II, Vienna 1961 and Vol.I and II, 1965
(7) H.H.Wickman, and G.K.Wertheim, in "Chemical Applications of the Mössbauer Effect", V.I.Goldanskii, and R.H. Herber, eds.,New York, Academic Press, 1968, p.604
(8) J.M.Friedt and J.Danon, Radiochim.Acta 17,(1972)
(9) A.G.Maddock in: "Nuclear Transformations in Solids", G.Harbottle and A.G. Maddock, eds., North-Holland Publishing Company, Amsterdam, in press
(10) N.N.Greenwood and T.G.Gibb, "Mössbauer Spectroscopy", Chapman and Hall, London 1971, p.329
(11) G.Kaindl, D.Kucheida, W.Potzel, F.E.Wagner, U. Zahn, and R.L. Mössbauer, in: "Hyperfine Interactions in Excited Nuclei", G.Goldring and R. Kalish, eds., Gordon and Breach,New York 1971, Vol.2, p.595

(12) W.Potzel, F.E.Wagner, U.Zahn, R.L.Mössbauer, and J.Danon, Z.Phys. 240,306(1970)
(13) H.D.Bartunik,W.Potzel,R.L.Mössbauer,and G. Kaindl, Z.Phys. 240, 1(1970)
(14) M.O. Faltens and D.A.Shirley, J.Chem.Phys. 534249 (1970)
(15) F.E.Wagner and U.Zahn, Z.Phys.233,1(1970)
(16) P.Rother,F.Wagner, and U.Zahn, Radiochim.Acta 11,203(1969)
(17) References in "Inorganic Synthesis",T.Moeller, ed., McGraw Hill, New York
(18) F.E.Wagner,D.Kucheida,H.Spieler,U.Zahn,and G.Kaindl, Z.Physik, to be published
(19) D.B.Rogers,R.D.Shannon,A.W.Sleight,and J.L.Gillson,Inorg.Chem.8,841(1969)
(20) H.U.Rahman,Phys.Rev.B3,729 (1971)
(21) E.A.Harris and K.S.Ingvesson,J.Phys.C1,990(1968)
(22) G.Lang and W.T.Oosterhuis,J.Chem.Phys.51,3608 (1969)
(23) F.E.Wagner and W.Potzel,in:"Hyperfine Interactions in Excited Nuclei",G.Goldring and R.Kalsih, eds., Gordon and Breach,New York 1971,p.681
(24) J.J.Davies and J.Owen, J.Phys.C2,1405(1969)
(25) G.Kaindl, private communication
(26) K.Leary and N.Bartlett, to be published
(27) C.H.W.Jones and J.L.Warren,J.Chem.Phys.53,1740 (1970)
(28) G.J.Perlow and M.R.Perlow,J.Chem.Phys.48,955 (1968)
(29) K.A.K.Lott and M.C.R.Symons,J.Chem.Soc.973(1960)
(30) J.M.Friedt and J.P.Adloff,Comptes Rend.246C, 1356(1967)
(31) J.M.Friedt,E.Baggio-Saitovich,and J.Danon,Chem. Phys.Lett.7,603(1970)
(32) A.Cruset and J.M.Friedt,phys.stat.sol.(b) 44, 633(1971)
(33) G.K.Wertheim,H.J.Guggenheim,and D.N.E.Buchanan, J.Chem.Phys.51,1931(1969)

THE APPLICATION OF MÖSSBAUER SPECTROSCOPY TO BIOCHEMISTRY

C. E. Johnson

Oliver Lodge Laboratory
University of Liverpool
Liverpool L69 3BX, England

1. INTRODUCTION

Many biological molecules contain iron. Since the Mössbauer Effect provides a powerful probe of the chemical state and the environment of iron atoms, it was logical to expect that it would be applied to the study of proteins and enzymes. Indeed the first measurements of the Mössbauer spectrum of Fe^{57} in haemin were reported by Gonser (1) at the International Conference on the Mössbauer Effect in Paris in 1961, only four years after the appearance of Mössbauer's papers on the discovery of the effect.

The electronic changes which take place during a biochemical reaction are generally centred around the iron atoms. Hence the iron plays a very important role in biology, and not only is it interesting to the physicist to be able to get an introduction to biological problems through the Mössbauer Effect, but the results may be of real value to the biochemist in trying to understand the nature of the reactions in which the molecule is involved in the living cell.

The main groups of biological molecules which contain iron at their active centres are shown in the following diagram:

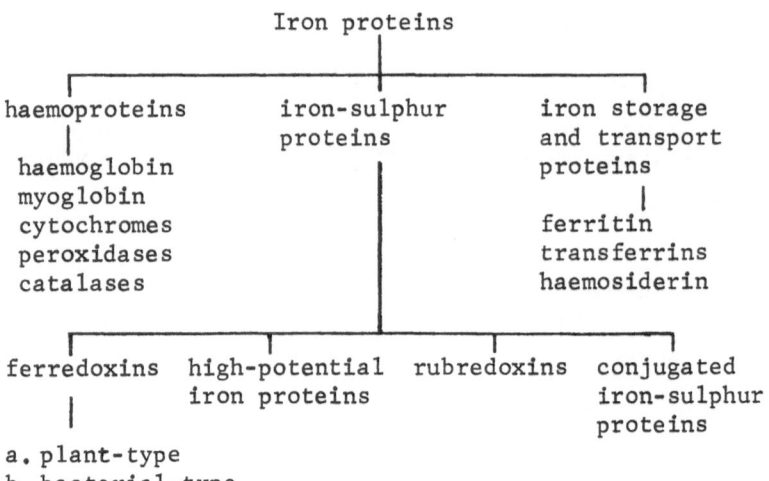

Iron proteins

- haemoproteins
  - haemoglobin
  - myoglobin
  - cytochromes
  - peroxidases
  - catalases
- iron-sulphur proteins
  - ferredoxins
    - a. plant-type
    - b. bacterial-type
  - high-potential iron proteins
  - rubredoxins
  - conjugated iron-sulphur proteins
- iron storage and transport proteins
  - ferritin
  - transferrins
  - haemosiderin

The haem proteins are the best understood of these molecules, and the first systematic study of biological molecules using the Mössbauer Effect was done on them by Lang and Marshall (2). Their molecular structure was known from X-ray diffraction measurements. The iron atoms are octahedrally co-ordinated to four nitrogen atoms in the haem group and to a fifth nitrogen in a histidine group which formed the link to the chain of amino-acids. The sixth ligand may be varied, and the state of the iron atom varies with it. Previous studies of the iron had been made by magnetic susceptibility and EPR techniques. In healthy blood the iron is ferrous, being low-spin when oxygenated and high-spin when de-oxygenated. Some ligands make the iron ferric, e.g. fluoride (high-spin) or cyanide (low-spin). The Mössbauer spectra of these molecules have been valuable in confirming these earlier conclusions, and have yielded quantitative data on the way that the energy levels and wavefunctions of the iron atoms are affected by the ligand field and spin-orbit coupling in the protein. They have also been valuable in providing standard spectra for each of the four common states of iron, and hence in establishing on a firm basis the use of Mössbauer spectroscopy as a way of obtaining information about iron in a biological (i.e., haem) environment. More recent work (3) has been directed towards the study of new haem derivatives, and the use of the Mössbauer spectrum to investigate the products of biochemical reactions. A review of work on haem proteins has been given by Lang (4).

Most of the rest of the Mössbauer work on biological molecules has been done on the iron-sulphur proteins. Unlike the haem proteins, little structural information is available on these molecules. The structure has only recently been determined for three of them. These are a rubredoxin, a ferredoxin (bacterial) - both from Clostridium Pasteurianum - and a high potential iron protein (HIPIP) from Chromatium. Although these molecules are different in size, structure and function, they all contain iron atoms in similar environments with the iron at the centre of four sulphur atoms which form an approximate tetrahedron. The simplest of these molecules are the rubredoxins, as they contain only one iron-sulphur group per molecule. Next in order of complexity are the plant ferredoxins and hydroxylase proteins (adrenodoxin and putidaredoxin), which contain two iron atoms per molecule which are close to each other and are strongly interacting magnetically. More complex still are the bacterial ferredoxins and HIPIPs; in these the iron appears to be in units containing four iron atoms which are strongly coupled together magnetically. Tsibris and Woody (5) have reviewed the physical data, including Mössbauer spectra, which have been obtained on the iron-sulphur proteins.

In this paper we shall describe the use of the Mössbauer Effect in the study of the iron-sulphur proteins. From a study of rubredoxin in the oxidized and reduced states, the chemical shifts and magnetic hyperfine interaction of Fe^{2+} and Fe^{3+} in tetrahedral sulphur co-ordination are measured. This effectively calibrates these quantities, i.e., it allows for the effects of covalency in this environment. These single-iron data are then used in the interpretation of the data on the two-iron proteins. Here the magnetic moments of the two iron atoms in the molecule are antiferromagnetically coupled together, and the resulting hyperfine spectrum is very different from those usually observed from a single iron atom. This is a situation unlike that found in inorganic complexes of iron, and it is not easy by any other method to observe directly the antiferromagnetic coupling between the pairs of iron atoms. Thus the Mössbauer Effect has been able to make a very real contribution to our knowledge of the iron-sulphur proteins.

The measurements are usually made on frozen aqueous

solutions of the proteins. The molecules are generally enriched in Fe^{57}, although some measurements have been done using naturally occurring iron. The Fe^{57} may be incorporated in the molecule in either of two ways; either by growing the organism from which the protein is extracted on the separated isotope Fe^{57}, or by incorporating it by chemical exchange. The growing method requires more Fe^{57}, but it is more reliable and more generally applicable than exchange, which must be tested carefully to ensure that the protein is not modified by the process.

2. CHEMICAL SHIFTS

Chemical shifts for several inorganic and biological molecules are given in the following table:

Ligands		Fe^{2+}		Fe^{3+}
6 H_2O	$FeSiF_6.6H_2O$	1.42	Fe_2O_3	0.50
4 O^{2-}	$FeBaSi_4O_{10}$	0.87	-	-
6 Cl^-	$FeCl_2$	1.20	$FeCl_3$	0.53
4 Cl^-	$(NMe_4)_2FeCl_4$	1.05	$(NMe_4)FeCl_4$	0.30
5N, H_2O or O_2	deoxyhaemoglobin	0.90	methaemoglobin	0.20
6 S^{2-}	-	-	Fe-tris-dtc	0.50
4 S^{2-}	rubredoxin	0.65	rubredoxin	0.25
	ferredoxin (plant)	0.56	ferredoxin (plant)	0.22
	adrenodoxin	-	adrenodoxin	0.26
	HIPIP	0.42	HIPIP	0.32

(shifts in mm/sec, measured at 77°K relative to iron metal at 290°K)

It is seen that the shift decreases as the degree of covalency of the ligands increases, i.e., in the order $-H_2O$, $-Cl^-$, $-O^{2-}$, $-N^{3-}$, $-S^{2-}$, $-CN^-$, etc. Also it is systematically less by about 0.2 mm/sec for tetrahedral co-ordination compared with octahedral co-ordination to the same ligands. It is clear that the chemical state cannot be inferred only from the value of the chemical shift alone. A possible exception might be high spin Fe^{2+} which usually has a large positive value of shift, but the Table shows that it can be as low as 0.6 mm/sec in tetrahedral sulphur co-ordination, which overlaps with the values found for Fe^{3+}.

Hence measurements of magnetic hyperfine coupling, which is very sensitive to the state of the iron atoms, are especially valuable in biological compounds.

3. MAGNETIC HFS IN IRON-SULPHUR PROTEINS

(i) <u>1-iron proteins</u> (rubredoxins)

(a) oxidized. The Mössbauer spectrum at 4.2°K shows a six-line Fe^{3+} pattern with an effective field H_n (Fe^{3+}) of -370 kG (6) (7). Since $g_{\parallel}$ for this state is 9.4, the hyperfine field H_n for Fe^{3+} is -395 kG. (In Fe_2O_3 it is -550 kG; in ferric tris-pyrrolidyldithiocarbamate, which has octahedral sulphur co-ordination, it is -475 kG.)

(b) reduced. The Mössbauer spectra are characteristic of Fe^{2+}, showing a large quadrupole splitting (3.3 mm/sec) and a shift which is low for Fe^{2+} which is consistent with tetrahedral sulphur ligands. Magnetic hyperfine splitting has been observed when large fields (up to 60 kG) are applied at 4.2°K. (9). The effective field at the nuclei is predominantly perpendicular to the symmetry axis of the ligand field, and so

$$\underline{H}_{eff} = \underline{H} + \underline{H}_{nx}.$$

In Figure 1 is plotted the internal field H_{nx} as a function of the applied field H. The saturated value of the hyperfine field H_{nx} (Fe^{2+}) is estimated to be -210 kG.

For comparison the internal fields observed (8) for

polycrystals of ferrous fluosilicate are also shown in Figure 1. This is a well understood salt, with H_{nx} (0) = -248 kG, and a ligand field splitting given by $D(S_z^2 - 2)$, where D = +10.9 cm^{-1}. Note that although the saturated value of the hyperfine field is smaller for the rubredoxin, the values for small applied fields are larger than for ferrous fluosilicate. Since the internal field is given for low applied fields ($g_\perp \beta H << D$) by

$$H_{nx} = \frac{3g_\perp \beta H}{D} H_{nx} (Fe^{2+})$$

a comparison with the fluosilicate data shows that in reduced rubredoxin, D = +4.4 cm^{-1}.

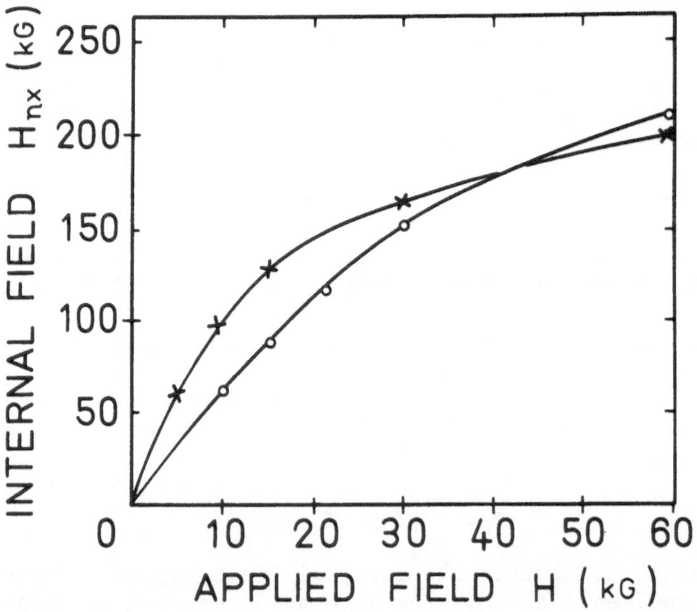

Figure 1. Internal magnetic fields in reduced rubredoxin (x) and in ferrous fluosilicate (o) as a function of the applied magnetic field.

The sign of the quadrupole coupling is seen to be negative, which shows that the ground state of the Fe^{2+} ion is d_{z^2}. The observed value for the hyperfine field may be understood using the approach of Marshall and Johnson (9). The core polarization field may be estimated to be 4/5 of the Fe^{3+} field, i.e., H_s = -315 kG. The dipolar field H_{dx} is about +64 kG. From the value of D, $g_\perp$ is estimated to be 2.04, and hence the orbital field H_{Lx} = +36 kG. This gives an expected total for H_{nx} (Fe^{2+}) of -215 kG.

(ii) 2-iron proteins (plant-type ferredoxins)

At 77°K the Mössbauer spectrum of the oxidized proteins consists of a doublet, while on reduction two doublets are observed, one due to Fe^{2+} and one due to Fe^{3+} (10). In both states the iron atoms are believed to be coupled together antiferromagnetically. When oxidized there are two Fe^{3+} atoms and their total spin is zero, which agrees with their temperature independent susceptibility. On reduction only one electron is transferred to the molecule, and the coupling between the Fe^{3+} atom (with S_1 = 5/2) and the Fe^{2+} atom (with S_2 = 2) results in a ground state with total spin $\underline{S} = \underline{S}_1 + \underline{S}_2 = 1/2$. This model was proposed to account for the unusual g-values centred around 1.94 found from the EPR spectra of these molecules (11).

In the reduced state the Mössbauer spectrum at low temperatures shows magnetic hyperfine splitting because the electron spin-lattice relaxation time has become long compared with $\hbar/a$, where A is the hyperfine coupling. The observed spectrum is asymmetrical and has broad and poorly resolved lines, but when a small magnetic field $H > AS/g\beta$ is applied a more symmetrical spectrum with sharper lines results. The effect of the field is to cause the electron spins to precess about it resulting in an effective magnetic field at the nuclei, rather than having the electronic and nuclear spins precessing together about their resultant.

Because of the coupling between the two atoms the effective field at their nuclei is reduced compared with that of the free atoms, and is proportional to the component of their spins along the total spin S, i.e.,

$$H_{1z} = \frac{\underline{S}_1 \cdot \underline{S}}{S_1(S+1)} H_{nz} (Fe^{3+}) = \frac{7}{15} H_{nz} (Fe^{3+})$$

$$H_{2z} = \frac{\underline{S}_2 \cdot \underline{S}}{S_2(S+1)} H_{nz} (Fe^{2+}) = -\frac{1}{3} H_{nz} (Fe^{2+})$$

For the Fe^{3+} atom H_n will be approximately isotropic, and taking the value found in oxidized rubredoxin, one would expect H_{1z} = -185 kG, in good agreement with the observed value of about -180 kG. For the Fe^{2+} ion the hyperfine

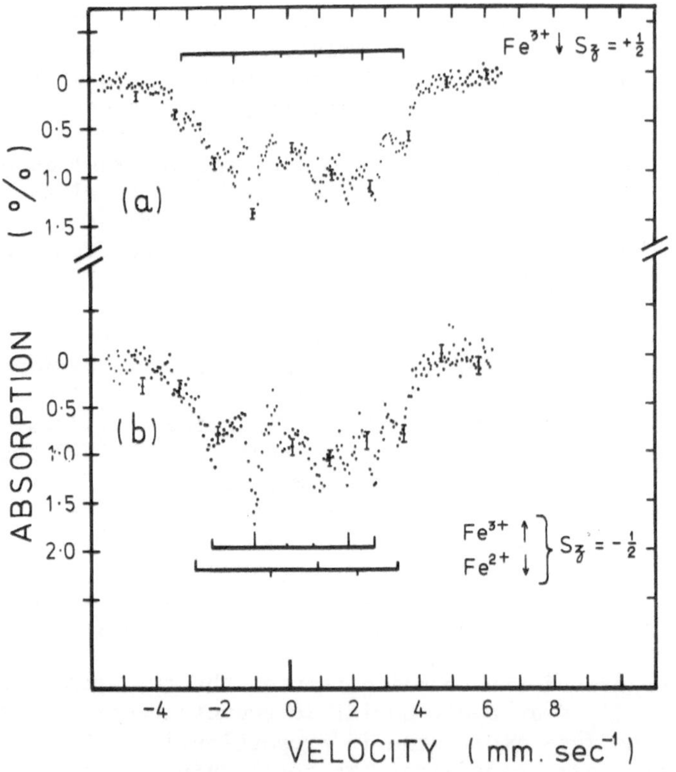

Figure 2. Mössbauer spectra of reduced spinach ferredoxin in 30 kG applied perpendicular to the γ-ray beam at (a) 4.2°K and (b) 1.7°K.

field is anisotropic. Taking the value for H_s found in reduced rubredoxin, and estimating H_{Lz} (+40 kG) and H_{dz} (-128 kG for the d_{z^2} orbital ground state), we expect H_{nz} (Fe^{2+}) to be -403 kG. Thus H_{2z} should be +135 kG, again agreeing with the measured value of about 120 kG.

On increasing the size of the applied field the spectrum splits into two components, showing that H_{1z} and H_{2z} have different signs owing to the antiferromagnetic coupling between the two iron atoms. This is shown in Figure 2 for spinach ferredoxin (10). At 1.7°K the population of the $S_z = +\frac{1}{2}$ state is small (9%) and the lines due to $Fe^{3+}\uparrow$ and $Fe^{2+}\downarrow$ can be separated out as shown. Although the use of stick diagrams is not strictly valid when the hyperfine interaction tensor is anisotropic, it has been shown that they can give a remarkably good approximation to the observed spectrum for many Fe^{2+} salts (see e. g. (8)). On warming the samples to 4.2°K, the $S_z = +\frac{1}{2}$ state becomes appreciably populated (28%) and contributes extra lines to the spectrum (see Figure 2 (a) compared with (b)).

Thus the Mössbauer spectrum allows the antiferromagnetic coupling between the Fe^{2+} and Fe^{3+} atoms in reduced 2-iron proteins to be confirmed, although it is not easy to identify all the Fe^{2+} lines from the Mössbauer spectrum above. However, the hyperfine interaction tensor for the Fe^{2+} atoms has been measured (12) in some of these proteins using ENDOR, and using these data the Mössbauer specta of spinach ferredoxin (13) and putidaredoxin (14) have been interpreted in detail.

Acknowledgements

I am indebted to my colleagues D. O. Hall, R. Cammack, K. K. Rao, M. C. W. Evans, and L. W. Becker for help and discussion and the Science Research Council for support.

REFERENCES

(1) U. Gonser, Proc. 2nd. Int. Conf. on Mössbauer Effect, Paris (Wiley 1962) p. 280.

(2) G. Lang and W. Marshall, Proc. Phys. Soc. 87, 3 (1966).

(3) M. R. C. Winter, C. E. Johnson, G. Lang, and R. J. P. Williams, Biochim. Biophys. Acta 263, 515 (1972).

(4) G. Lang, Quart. Rev. Biophys. 3, (1970).

(5) J. C. M. Tsibris and R. W. Woody, Co-ordination Chemistry Reviews (Elsevier 1969).

(6) W. D. Phillips, M. Poe, J. F. Weiher, C. C. McDonald, and W. Lovenberg, Nature 227, 574 (1970).

(7) K. K. Rao, M. C. W. Evans, R. Cammack, D. O. Hall, C. L. Thompson, P. J. Jackson, and C. E. Johnson, Biochem. J. (1972) in press.

(8) C. E. Johnson, Proc. Phys. Soc. 92, 748 (1967).

(9) W. Marshall and C. E. Johnson, J. Phys. Rad. 23, 733 (1962).

(10) K. K. Rao, R. Cammack, D. O. Hall, and C. E. Johnson, Biochem. J. 122, 257 (1971).

(11) J. F. Gibson, D. O. Hall, J. H. M. Thornley, and F. R. Whatley, Proc. Nat. Acad. Sci. U. S. 56, 987 (1966).

(12) J. Fritz et al., Biochim. Biophys. Acta 253, 110 (1971).

(13) W. R. Dunham, et al., Biochim. Biophys. Acta 253, 134 (1971).

(14) E. Munck, P. G. Debrunner, J. C. M. Tsibris, and I. C. Gunsalus, Biochem. 11, 855 (1972).

MÖSSBAUER STUDIES OF AN ENZYME SYSTEM: PUTIDAREDOXIN AND CYTOCHROME P450

Peter G. Debrunner

Physics Department, University of Illinois at Urbana-Champaign, Urbana, Illinois 61801

Iron proteins play a vital role in many biochemical reactions. Hemoglobin and myoglobin for instance bind molecular oxygen, cytochrome c is active in electron transport, and other iron proteins catalyze specific biochemical reactions. Cytochrome P450, the main topic of this discussion, combines all these features: It binds molecular oxygen and inserts one atom of it into a specific organic molecule. The reaction requires the transfer of two electrons, one at a time, through a specific iron-sulfur protein, putidaredoxin. Fig. 1 shows the complete enzymatic cycle(1) which will be discussed in more detail below and Table 1 indicates the wide occurrence and the biological significance of the cytochrome P450 enzyme system.

^{57}Fe Mössbauer spectroscopy provides a sensitive tool for probing the environment of the iron and, since much of the enzymatic action in these proteins takes place near the iron, it can yield information about certain aspects of the reaction mechanism. It is clear that a Mössbauer measurement yields a static picture only. Kinetic information, which is equally important for an understanding of the reaction mechanism, has to come from other types of experiments.

It is possible, though, to study a reaction, like the one in Fig. 1, in slow motion by taking Mössbauer spectra of the reaction intermediates at various stages along the cycle.

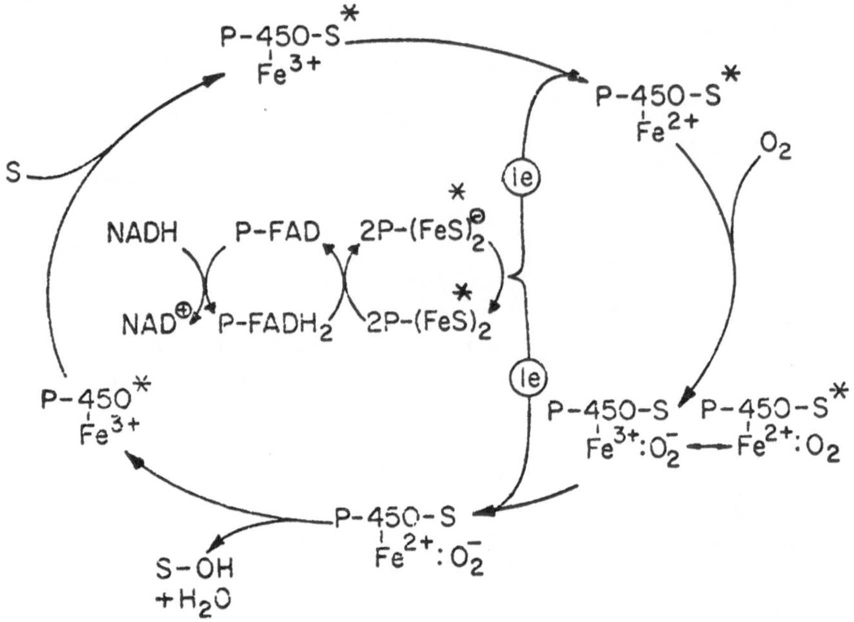

Fig. 1 Proposed reaction mechanism for the hydroxylation of camphor in Pseudomonas putida. NADH (NAD^+) is a biological reducing agent in its reduced (oxidized) state, P-FAD is a flavoprotein, P-$(FeS)_2$ is the 2Fe-2S protein, putidaredoxin, S and S-OH stand for the substrate, D-camphor, and the product, hydroxylated camphor, respectively. Species marked with an asterisk have been studied by Mössbauer spectroscopy and are discussed in this paper.

The enzyme system of Fig. 1, which hydroxylates camphor, has been isolated from a bacterium, Pseudomonas putida, by Gunsalus and coworkers. (2)

TABLE I TYPICAL P450 HYDROXYLASE SYSTEMS

(Biological Reductant) (Flavo-Protein) (2 Fe-2 S Protein) (Cytochrome P450) (H_2O+Product / O_2+Substrate)

Such systems are found in

Bacteria

e.g. Camphor Hydroxylase of Pseudomonas Putida[a)]
NADH - Flavoprotein - Putidaredoxin - $P450_{cam}$ - O_2 + Camphor

Adrenal Mitochondria

e.g. Steroid 11-β Hydroxylation[b)]
NADPH - Flavoprotein - Adrenodoxin - P450 - O_2 + Substrate
or Cholesterol Side Chain Cleavage[c)]

Liver Microsomes (No Fe-S Protein)

e.g. ω-Hydroxylation[d)]
Drug Detoxification[b)]

---

a) M. Katagiri, B. N. Ganguli and I. C. Gunsalus, J. Biol. Chem. 243, 3543 (1968).

b) T. Omura, R. Sato, D. Y. Cooper, O. Rosenthal and R. W. Estabrook, Federation Proc. 24, 1181 (1966).

c) E. R. Simpson and G. S. Boyd, Biochem. Biophys. Res. Commun. 24, 10, 1966).

d) A. Y. H. Lu and M. J. Coon, J. Biol. Chem. 243, 1331 (1968).

Similar enzyme systems are found in mammalian cells (Table 1). Because of their important functions, e.g. steroid metabolism and drug detoxification, great efforts are made in their investigation. The Pseudomonas hydroxylase is very useful as a model system; it offers great advantages for experimentation over the mammalian system because it is soluble, in contrast to most of the others, and because it is available in large quantity and high purity.

The camphor hydroxylase consists of three proteins, two of which contain iron. Mössbauer measurements on these two proteins, putidaredoxin and cytochrome $P450_{cam}$, will be discussed below.

Putidaredoxin is a relatively small protein of molecular weight, 12,500, which contains two iron atoms per molecule. The iron is bound directly to the protein, presumably through four of the cystein residues. At low pH the iron is released together with two sulfide ions, but the protein can again be reconstituted at higher pH. Isotopic substitution of the iron and sulfur can therefore readily be achieved. Putidaredoxin exists in an oxidized, diamagnetic state and in a one-electron reduced state of spin $S = \frac{1}{2}$, which shows a characteristic ESR signal with $g_\perp = 1.94$, $g_\parallel = 2.01$. ESR studies(3) on ^{33}S and ^{57}Fe enriched samples have shown that the unpaired spin interacts with the two iron and labile sulfur atoms as well as with sulfur on the protein. Many other 2Fe-2S proteins are known that closely resemble putidaredoxin in its physical and chemical properties,(4) the closest relative being the adrenodoxin listed in Table 1.

On the basis of much more limited information Gibson et al.(5) in 1966 proposed a model of the 2Fe-2S complex which was borne out by Mössbauer data (Fig. 2) of putidaredoxin(6) and similar experiments by other groups.(7-9) According to Gibson's model the two iron atoms a and b are antiferromagnetically coupled. In the oxidized protein they are both in the ferric high-spin state and the two spins, $S_a = S_b = 5/2$, couple to a resultant spin zero. In the reduced protein, however, one of the irons, say b, has changed to the high-spin ferrous state, $S_b = 2$, and the pair couples to a total spin of $S = 1/2$. In order to explain the g-tensor of the reduced protein Gibson further assumed distorted tetrahedral symmetry at the b-site. This assumption is again compatible with the Mössbauer data, but other symmetries can not be excluded. It is tempting to assume that both irons are tetrahedrally coordinated to four sulfur atoms such that the two labile sulfurs act as bridging ligands between the two iron atoms, allowing Fe-S-Fe superexchange, whereas the remaining four sulfur ligands are provided by the four cysteins that are invariably found in the 2Fe-2S proteins. More work is clearly required to settle this question.

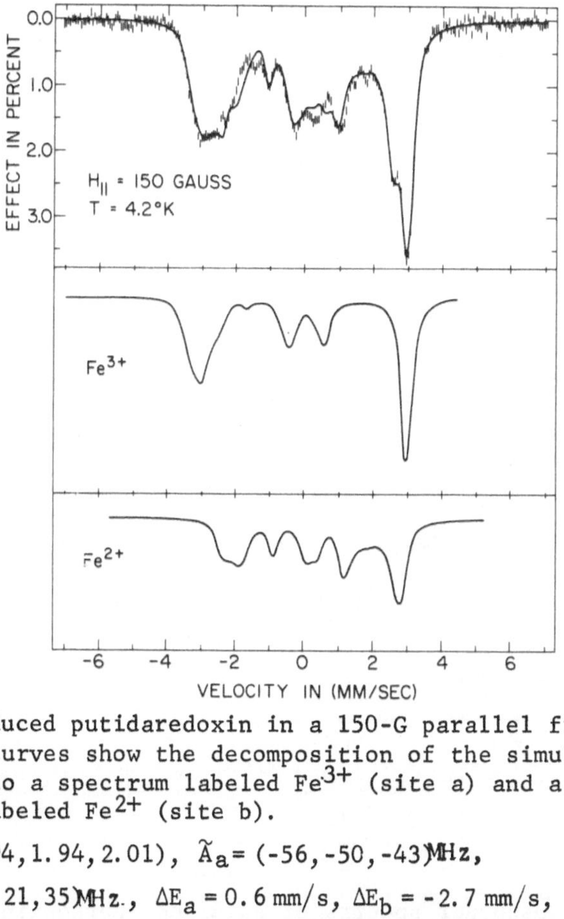

Fig. 2 Reduced putidaredoxin in a 150-G parallel field. The lower curves show the decomposition of the simulated spectra into a spectrum labeled Fe^{3+} (site a) and a spectrum labeled Fe^{2+} (site b).

$$[\tilde{g} = (1.94, 1.94, 2.01),\ \tilde{A}_a = (-56, -50, -43) \text{MHz},$$
$$A_b = (14, 21, 35) \text{MHz},\ \Delta E_a = 0.6 \text{ mm/s},\ \Delta E_b = -2.7 \text{ mm/s},$$
$$\eta_a = 0.5,\ \eta_b = -3,\ \delta_a - \delta_b = -0.31 \text{ mm/s},\ \Gamma = 0.31 \text{ mm/s}].$$

Fig. 3 shows Mössbauer spectra of oxidized rubredoxin, (10) a related iron-sulfur protein of known structure (11) with a single iron atom bound to four cystein ligands. We are presently analyzing these spectra to have a basis for comparison with the hyperfine parameters of putidaredoxin.

Cytochrome $P450_{cam}$ has a molecular weight of 45,000 and contains one heme group and thus one atom of iron per molecule. The enzyme was prepared from cultures grown on an ^{57}Fe-enriched medium(12) and showed an enrichment of more than 80%. Typical Mössbauer samples contained 0.5 to 1 mℓ of a frozen solution of approximately 2mM $P450_{cam}$ in 50 mM potassium phosphate buffer, pH7.

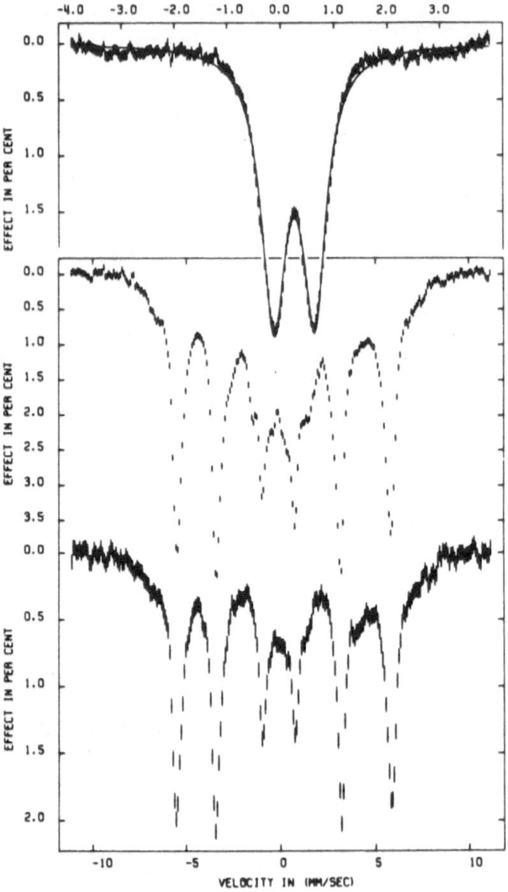

Fig. 3 Mössbauer spectra of oxidized rubredoxin in a 100-G parallel field at 77 K (top), and in a 1.3-kG perpendicular field at 4.2 K (middle) and 1.2 K (bottom).

Cytochrome $P450_{cam}$ forms a number of relatively stable complexes which have been characterized as reaction intermediates in the enzymatic cycle of Fig. 1. The four complexes marked with asterisks have been studied by Mössbauer spectroscopy.(13) In addition we have measured the CO adduct of reduced $P450_{cam}$; this species shows the distinctive optical absorption peak at $\lambda = 450$ nm for which the P450 cytochromes are named. Representative Mössbauer spectra are shown in Figures 4 and 5.

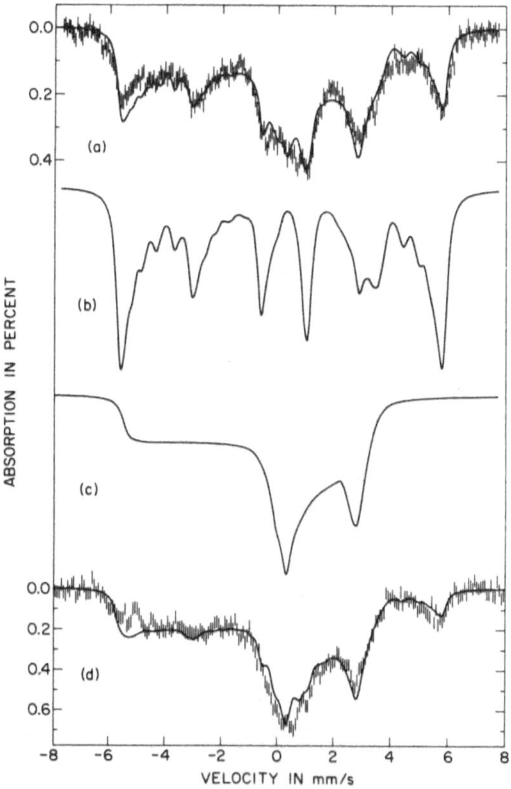

Fig. 4 Oxidized $P450_{cam}$ at 4.2 K in a 120-G parallel field (a) with camphor (>3.5mM), (d) camphor-depleted. Curve (b) is a simulation of the high-spin component [$\tilde{g}$=(8,4,1.8), $\tilde{A}$=(-8.53,-4.27,-1.92)mm/s, $e^2qQ/2$=0.77mm/s, η=0.63, δ_{Fe}=0.43]. Curve (c) is a simulation of the low-spin component [$\tilde{g}$=(2.45, 2.26,1.91), $\tilde{A}$=(2.39,0.98-5.41)mm/s, $e^2qQ/2$=-3.3mm/s, η=0, δ_{Fe}=0.06mm/s]. The solid curves superimposed on the data are sums of (b) and (c) in the ratio 45:55 in (a) and 20:80 in (d).

In the interpretation of the $P450_{cam}$ data we are on much safer grounds than in the previously discussed case of putidaredoxin because we know the structure of the heme group and we have a vast amount of literature available, both experimental and theoretical, on other heme proteins

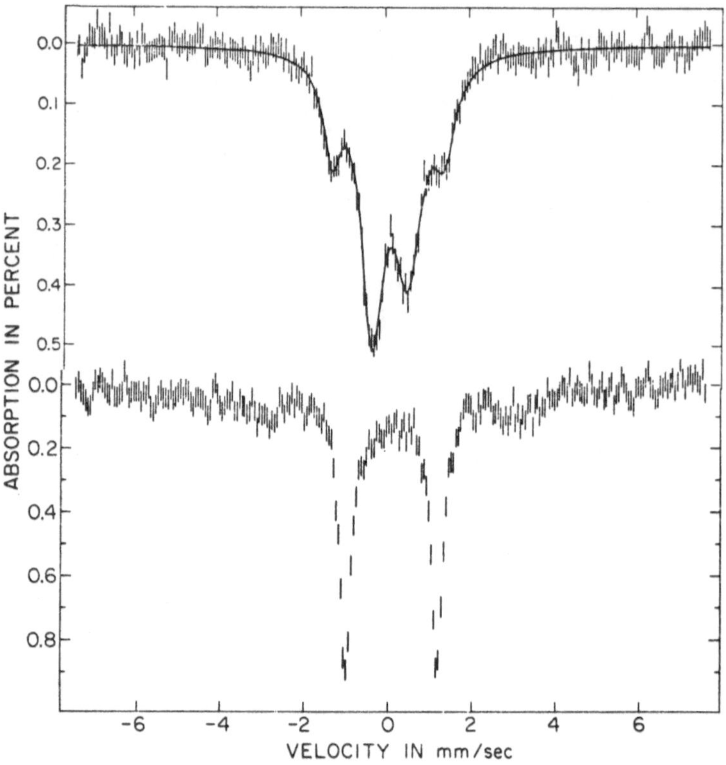

Fig. 5 Upper curve: Oxidized $P450_{cam}$ in >3.5mM camphor at 200 K. The solid line represents a least squares fit with two symmetric doublets. The inner (outer) doublet with ΔE_Q=0.79mm/s, δ_{Fe}=0.35mm/s (ΔE_Q=2.66mm/s, δ_{Fe}=0.31mm/s) represents high-spin (low-spin) iron.

Lower curve: Oxygenated $P450_{cam}$ at 4.2 K in a 1.3-kG perpendicular field. The quadrupole doublet with ΔE_Q=2.15 $\pm$ 0.02mm/s and δ_{Fe}=0.31 $\pm$ 0.02mm/s accounts for 70% of the total area and represents the O_2-complex. The remaining 30% of the area are due to oxidized $P450_{cam}$.

and model compounds. In addition, optical(14) and magnetic resonance(15) data on $P450_{cam}$ have already been published.

In the heme group the iron is invariably bound to the four nitrogen atoms of the porphyrin ring. To lowest approximation the group is planar and has 4-fold symmetry about an axis perpendicular to the heme plane. The iron is normally six-coordinated, in certain cases only five-coordinated, i.e. it has axial ligands above and below the heme plane. Depending on the strength and the symmetry of these ligands relative to the porphyrin the iron can exist in a variety of charge and spin states. It is believed that this variability and the possibility of electron transfer to and from the porphyrin ring account for the reactivity of the heme group. (16)

The P450 cytochromes are unusual heme proteins in many respects. The distinctive optical absorption of the CO complex may be an accidental feature, but it may also indicate a characteristic ligand. The magnetic properties of high-spin ferric $P450_{cam}$, i.e. the g-and A-tensors, deviate strongly from tetragonal symmetry; in a crystal field model a large rhombic distortion is required. (15) At present none of the axial ligands is known with certainty.

In camphor-free oxidated $P450_{cam}$ the iron is present in the low spin ferric state, $Fe^{3+}, S = 1/2$ (Fig. 4d). Upon addition of the substrate, camphor, $P450_{cam}$ forms a 1:1 camphor-complex and the iron partially changes to its high spin state, $S = 5/2$ (Fig. 4a). Camphor presumably binds to the protein near the heme group. Whether a conformational change of the protein or only a rearrangement of the axial heme ligand is involved is not known. As shown by Lang(17) and others the Mössbauer spectra of ferric iron compounds in the presence of static hyperfine interaction can be calculated from the spin Hamiltonian

$$\mathcal{H} = \beta \vec{S}\tilde{g}\vec{H} + \vec{S}\tilde{A}\vec{I} + \frac{e^2qQ}{4I(2I-1)}\left[3I_z^2 - I(I+1) + \eta(I_x^2 - I_y^2)\right].$$

We have used the Griffith model(18) to calculate the A- and EFG-tensors for low-spin $P450_{cam}$ from the known g-tensor. (15) The Mössbauer spectra simulated with these parameters fit the data reasonably well; the solid lines in Fig. 4c represent an improved fit obtained by modifying the parameters slightly.

To simulate the spectra of the high-spin species (Fig. 4b) we have used the spin Hamiltonian above, the g-tensor as measured by ESR, (15) the relation $A_i = \mathrm{const} \cdot g_i$, $i = 1, 2, 3$ and the quadrupole splitting measured at 200 K. The Mössbauer parameters of the reduced camphor complex ($\Delta E_Q = 2.45$ mm/s, $\delta_{Fe} = 0.83$ mm/s at 4.2 K) are typical of high-spin ferrous iron, S=2. The spectrum of the oxygen adduct (Fig. 5, bottom) shows a striking similarity with the spectrum of oxyhemoglobin. (17) A high-field measurement proves that oxy-$P450_{cam}$ is diamagnetic. The unusual combination of small positive isomer shift, relatively large quadrupole splitting and diamagnetism appears to be a characteristic feature of the heme-oxygen adduct; it is most naturally explained as an antiferromagnetically coupled complex of low-spin ferric iron (Fe^{3+}, S = 1/2) with a superoxide anion (O_2^-, S = 1/2).

The results presented here are preliminary in many respects and we want to improve and expand them. The main task, however, will be to find models that can explain our data and give us a better understanding of the reaction mechanism.

REFERENCES

1. I. C. Gunsalus, C. A. Tyson, R. Tsai and J. D. Lipscomb, Chem.-Biol. Interactions 4, 75 (1971).
2. I. Hedegaard and I. C. Gunsalus, J. Biol. Chem. 240, 4038 (1965).
3. J. C. M. Tsibris, R. Tsai, I. C. Gunsalus, W. H. Orme-Johnson, R. E. Hansen and H. Beinert, Proc. Nat. Acad. Sci. U. S. 59, 959 (1968).
4. J. C. M. Tsibris and R. W. Woody, Coordin. Chem. Rev. 5, 417 (1970).
5. J. F Gibson, D. O. Hall, J. H. M. Thornley, and F. R. Whatley, Proc. Nat. Acad. Sci. U. S. 56, 987 (1966).
6. E. Münck, P. G. Debrunner, J. C. M. Tsibris, and I. C. Gunsalus, Biochemistry 11, 855 (1971).
7. W. R. Dunham, A. Bearden, I. Salmeen, J. Fee, D. Petering, R. H. Sands and W. H. Orme-Johnson, Biochim. Biophys. Acta 253, 134 (1971).
8. C. E. Johnson, R. Cammack, K. K. Rao and D. O. Hall, Biochem. Biophys. Res. Commun. 43, 564 (1971).
9. K. K. Rao, R. Cammack, D. O. Hall and C. E. Johnson, Biochem. J. 122, 257 (1971).
10. The sample was prepared by Dr. Lovenberg.
11. J. R. Herriott, L. C. Sieker, L. H. Jensen and W. Lovenberg, J. Mol. Biol. 50, 391 (1970).

12. The enzyme was prepared and tested by V. Marshall and J. D. Lipscomb.
13. M. Sharrock, E. Münck, P. G. Debrunner, V. Marshall, J. D. Lipscomb and I. C. Gunsalus, submitted to Biochemistry.
14. See e. g. I. C. Gunsalus, C. A. Tyson, R. Tsai and J. D. Lipscomb, Chem. -Biol. Interactions 4, 75 (1971).
15. R. Tsai, C. A. Yu, I. C. Gunsalus, J. Peisach, W. Blumberg, W. H. Orme-Johnson and H. Beinert, Proc. Nat. Acad. Sci. U. S. 66, 1157 (1970).
16. In this context it is interesting to mention Mössbauer emission experiments which show that the heme group survives radioactive decay of either an axial ligand or of the central metal ion.
17. G. Lang and W. Marshall, Proc. Phys. Soc. 87, 3 (1966).
18. J. S. Griffith, Nature 180, 30 (1957).

INTERPRETATION OF MÖSSBAUER ISOMER SHIFTS AND QUADRUPOLE SPLITTINGS OF FERROUS HEMOGLOBINS BY LIGAND FIELD AND MOLECULAR ORBITAL THEORY [+]

A. Trautwein [++]

Angewandte Physik
Universität des Saarlandes
66 Saarbrücken 11, West Germany

Dedicated to U. Gonser for his 50th birthday.

[+] Supported in part by U.S. National Science Foundation Grant GP-31373K, in part by an award from the Biomedical Sciences Support Grant at the University of Utah (U.S. Public Health Service Grant RR 07092), in part by the Centre Européen de Calcul Atomique et Moléculaire, Orsay, France, and in part by the Stiftung Volkswagenwerk.

[++] Supported at the Centre Européen de Calcul Atomique et Moléculaire, Orsay, France by a fellowship (No. 531) of the European Molecular Biology Organization, and supported by the Deutsche Forschungsgemeinschaft (AZ 77/732/71) for presenting this work at the International Conference on Application of the Mössbauer Effect, Israel, 28.-31.8.1972.

Introduction

The study of the ferrous heme compounds carbomonoxyhemoglobin (HbCO), deoxyhemoglobin (Hb), and oxyhemoglobin (HbO) which are subject to the present investigation was one of the first applications of Mössbauer spectroscopy in biophysics (Gonser et al /1,3/, Karger /2/, Lang et al /4/).

From these and later publications (Kappler et al /5/, Maling et al /6/) the experimental results for HbCO, Hb, and HbO_2 as shown in fig. 1 and table I are well established.

The interpretation of experimental quadrupole splittings, ΔE_Q, of Hb was first based on electric field gradient (EFG) contributions from Fe 3d single electron orbitals wighted by an appropriate Boltzmann factor /3,5/. However, it turned out that a single electron in a Fe $3d_{xy}$ orbital would produce a ΔE_Q -value which is nearly twice of the experimental value measured for Hb at 4.2 K. Therefore spin-orbit coupling between low lying many electron terms $^5B_2$, ^5E, ^3E, and $^1A_1$ was taken into account to mix negative EFG contributions of ^5E and ^3E into the ground-state $^5B_2$, resulting in a calculated ΔE_Q -value which was comparable with experimental data (Eicher et al /7,8/). Corresponding calculations were carried out for the isomer shift δ of Hb (Trautwein et al /9/).

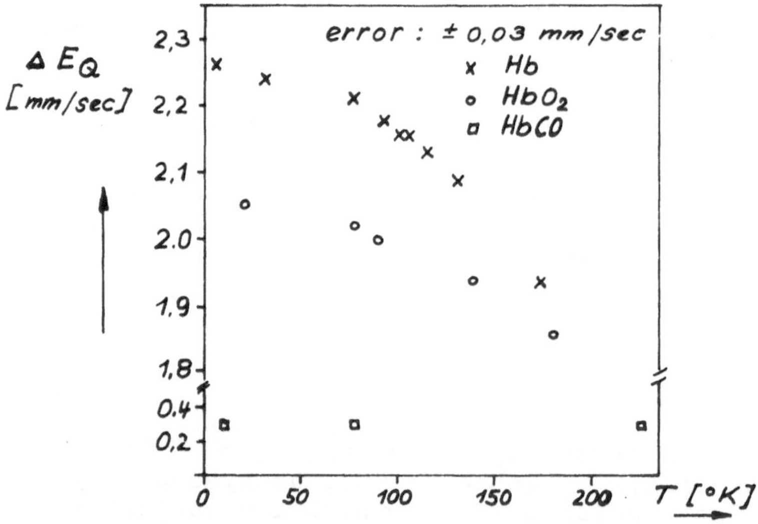

Fig. 1 Temperature dependence of experimental quadrupole splittings of HbCO, Hb, and HbO_2 . The absorbers are ^{57}Fe enriched rat hemoglobin, ph 10.

Table I

Typical experimental isomer shifts δ (relative to α-Fe) and quadrupole splittings ΔE_Q for HbCO, Hb, and HbO_2.

compound	T(K)	δ (mm/sec)	ΔE_Q (mm/sec)	sign EFG
HbCO	4.2	0.3	0.35	+a
Hb	4.2	0.95	2.25	+
HbO_2	4.2	0.3	2.05	-

(a) taken from reference /18/; obtained by fitting a magnetic hyperfinespectrum of carbomonoxy-myoglobin.

This ligand field type approach was also applied to HbCO (Eicher et al /7/, Trautwein et al /10/), resulting in an energetically isolated groundstate $^1A_1$.

For HbO_2 possible electronic groundstates which are in agreement with sign and absolute value of the experimental EFG were suggested (Trautwein et al /11/) for the coplanar bonding of O_2 to Hb (Griffith model /12/) and for the bent end-on structure of O_2 (Pauling model /13/) using the ligand field results for Hb and triplet states of O_2.

A qualitative interpretation of the oxyhemoglobin quadrupole splitting was given (Lang /14/) on the basis of covalency effects in the Fe d_π subshell due to the Griffith geometry or as an alternative on the basis of a partial population of the Fe $3d_{3z^2-r^2}$ orbital due to the coaxial bonding of O_2 to hemoglobin (old Pauling model which is believed to generate ferric iron /15/). The temperature dependence of ΔE_Q(T) was suggested to result from rotation of O_2 or thermally induced vibration of O_2 and/or adjacent parts of the protein /14/.

The interpretation of quadrupole splittings and isomer shifts in hemecompounds on the basis of MO calculations (Zerner et al /16/) was given by Weissbluth et al /17/ and is attempted by Trautwein et al /18/.

In the present work ligand field calculations on Hb and HbCO /7-10/ are reviewed and MO calculations on HbCO and HbO_2 (Trautwein et al /18-21/) are described with the goal to interpret experimental ΔE_Q and δ data of HbCO, Hb, and HbO_2 , and to give a possible pathway for the reaction of deoxyhemoglobin with oxygen.

Ligand Field Theory

1.) General. In Fe porphyrin complexes the symmetry of the environment of the central iron is almost C_{4v} . The iron is coordinated by a square array of pyrrole nitrogens plus certain atoms or molecules at one or both out-of-plane (fifth and sixth) positions on the iron, shown in fig. 2 for Hb, HbO_2 and HbCO.

The first aim is to evaluate within the configuration $3d^6$ (admixtures from configuration $3d^5 4p^1$ may be neglected /22/) the eigenvalues and eigenvectors of a Hamiltonian, which takes into account the proper point symmetry of the iron cation and the Coulomb repulsion of Fe 3d-electrons. For reasonable single electron energies the many electron terms $^5B_2$, ^5E, ^3E, and $^1A_1$ are found to be low in energy. Therefore spin-orbit coupling between these many electron wavefunctions was included for calculating ΔE_Q and δ . A general covalency factor α^2 of about 0.8 takes care of

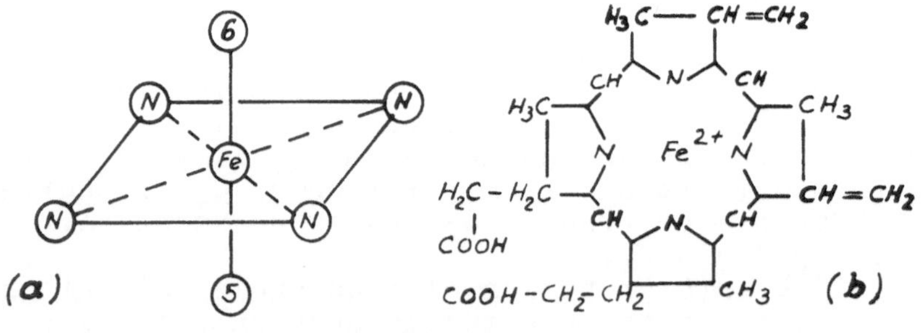

Fig. 2 (a) Iron surrounded by six (HbCO, HbO_2) or five (Hb) next neighbors.
(b) Structure of hemegroup.

correcting the free ion spin-orbit coupling constand λ and the EFG parameter $\langle r^{-3} \rangle$. The neglect of lattice contributions to the EFG is because of nearly zero ligand charges /16,18/ a reasonable approximation. The solution of the secular problem for the low lying levels $^5B_2$, ^5E, ^3E, and $^1A_1$ is expressed by the single electron energies ε_i. The 22 eigenvectors $|e_\mu\rangle$ are certain linear combinations of base vectors of the $3d^6$ configuration, each contributing to $\Delta E_{Q,total}$ and δ_{total} by typical high spin, intermediate spin, and low spin values. In order to evaluate the temperature dependence of ΔE_Q it is necessary to average the EFG components $V_{ZZ}(\mu)$ of the individual substates $|e_\mu\rangle$ according to Boltzmann statistics /23/.

2.) Hb. Applying this ligand field model to Hb suggests that the observed isomer shift δ (4.2 K) and temperature dependence ΔE_Q (T) can only be accounted for if the ^3E term is relatively close in energy to the groundstate $^5B_2$. Figure 3 shows fit curves of $\Delta E_{Q,Hb}$ (T) for typical energies of the many electron terms involved; furthermore the experimental and calculated values of δ (4.2 K) coincide within the experimental range of error for these energy-sequences. Changes of the covalency between 0.6 and 0.9 affect the energy of the ^3E term only within a range of about 10 cm^{-1}.

By adding further more or less open parameters to the model-like different covalency factors α_i^2 for the various Fe 3d orbitals - it might be possible to fit $\Delta E_{Q,Hb}$ (T) by taking into account only high spin terms $^5B_2$ and ^5E; however, the calculated isomershift would be in gross disagreement with the experimental value.

3.) HbCO. The small and temperature - independent quadrupole splitting observed for HbCO (table I) can be explained in terms of our ligand field approach only by an isolated spin singlet $^1A_1$ as groundstate, otherwise excited electron states could mix into the groundstate via spin-orbit coupling. An ^3E term lying above the $^1A_1$ term at a distance of about 1000 cm^{-1} mixes EFG contributions into the groundstate resulting in a quadrupole splitting of approximately 0.65 mm/sec. This leads to the conclusion that in HbCO the terms $^5B_2$, ^5E, and ^3E must lie at least at a distance of 2000 cm^{-1} above the groundstate $^1A_1$.

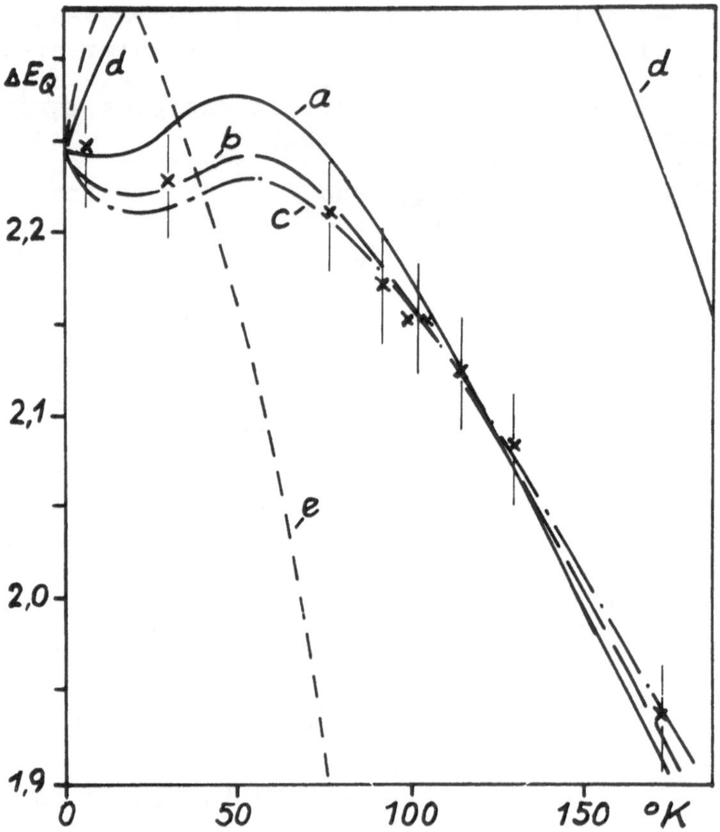

Fig. 3 Computer results for the quadrupole splitting (in mm/sec) in Hb plotted as a function of temperature. The rhombic perturbation parameter of ^5E is D=100 cm^{-1}. The energies relative to the groundstate $^5B_2$ are for E (^5E), E(^3E), and E($^1A_1$) respectively, in cm^{-1}.

Curve a 1500, 268, 2150
b 2000, 257, 2150
c 2500, 251, 2150, and for comparison
d 250,1750, 2150
e 347, 275, 1390

Molecular Orbital Theory

1.) General. Oxy- and carbomonoxyhemoglobin are believed to involve to a considerable extent covalent binding. In such cases it is useful to replace ligand field theory by molecular orbital theory. For the present study an approximate MO calculation was worked out and applied to a variety of iron- and non-iron-complexes (totally about 60) /18,19,20,21,24/. The method is based on an Iterative Extended Hückel (IEH) approach. The iterative feature gives the calculation self consistent field character, and "extended" indicates the application to all valence electrons: H 1s, C 2s and $2p_{\pm1,0}$, N 2s and $2p_{\pm1,0}$, O 2s and $2p_{\pm1,0}$ and Fe $3d_{\pm2,\pm1,0}$, 4s and $4p_{\pm1,0}$. Our IEH program is not symmetry restricted in contrast to the program reported by Zerner et al /16/.
We may estimate isomer shifts from our MO calculations using the approximation that atomic orbitals on centers other than Fe have no density at the Fe nucleus. As the Fe 3d and 4p orbitals also vanish at the Fe nucleus, the only contributing atomic orbital in our calculations is Fe 4s. However, the Fe inner-shell densities are also affected by the molecular structure and at least the Fe 3s contribution cannot be neglected. We make use of the following ideas for evaluating $\rho(0)$, the electron density at the Fe nucleus:

1. Fe 1s- and Fe 2s-contributions to $\rho(0)$ remain constant within different Fe-compounds.
2. The Fe 3s-contribution to $\rho(0)$ is estimated by considering its lack of orthogonality to the valence orbitals of neighboring ligands. This nonorthogonality can be interpreted as requiring a renormalization of the Fe 3s-wave function /25/, with a consequent change in the coefficient of $|\psi_{3s}(0)|^2$ away from 2, even though the Fe 3s orbital is assumed to remain fully occupied. This effect can be calculated by orthogonalizing the atomic wave function ψ_{3s} to the occupied MO's ϕ_i. An estimate of the Fe 3d shielding effects on $|\psi_{3s}(0)|^2$ is based on HFAO values for various electronic configurations of Fe and its ions /26/.
3. The Fe 4s-contribution to $\rho(0)$ is given by the

product of $|\psi_{4s}(0)|^2$ and the amount of 4s occupancy. The latter is assumed to be given by the bond order value $P_{4s,4s}$. The value $|\psi_{4s}(0)|^2$ is taken from non relativistic HF calculations /27/ corrected by a relativistic correction term S'(Z) /28/.

4. We assume that overlap integrals are to be evaluated using HFAO's /29/, as Slater type orbitals (STO's) give qualitatively incorrect overlaps with Fe 3s.

This model for calculating $\rho(0)$ was applied so far to the low spin compounds $Fe(CN)_5NO^{-2}$, $Fe(CN)_6^{-3}$, and $Fe(CN)_6^{-4}$ /19/, and by extending our SCF calculations by a spin-projected Configuration Interaction (CI) calculation it was also applied to the high spin compounds FeF_6^{-3}, FeF_6^{-4}, FeF_6^{-5} /19,20/, and Fe_2 (iron dimers in solid noble gases) /21/. Comparing $\Delta\rho(0)$ values with corresponding experimental $\Delta\delta$ data we find for the so called isomer shift calibration constant $\alpha = \Delta\delta/\Delta\rho(0)$ a value of $-0.33 \pm 0.05\ a_0^3$ mm/sec.

The quadrupole splitting ΔE_Q is derived from equation (1):

$$\Delta E_Q = (1/2)e^2Q(1-R)\langle r^{-3}\rangle \left(\sum_{3d} P_{3d,3d}f_{3d} + \sum_{4p} P_{4p,4p}f_{4p}\right),$$

with $(1/2)e^2Q(1-R)\langle r^{-3}\rangle 4/7 \approx 3.5$ mm/sec for nearly neutral iron.

Lattice contributions are neglected as the next nearest neighbors of iron in the complexes under study here are nearly neutral. For the HbCO geometries used here the EFG asymmetry parameter vanishes. The 3d and 4p summations are over the appropriate sets of Fe orbitals. Angular factors are indicated as f_{3d} and f_{4p}. Equation (1), including lattice contributions, was applied to the compounds $Fe(CN)_5NO^{-2}$, $Fe(CO)_5$ /19/, and Fe_2 /21/ where satisfactory agreement between calculated and experimental quadrupole splittings was obtained.

2.) HbCO. In table II the molecular geometries, quadrupole splittings, and isomershifts for HbCO are summarized. The hemegeometries are denoted by ZGK and K, namely Zerner's et al /16/ planar and Koenig's /30/ non-planar hemegeometry. In some geometries a water molecule was added to the complex as indicated in fig. 4.

Table II

Molecular geometries, related isomershifts δ and quadrupole splittings ΔE_Q for HbCO.

No. of geometry	1	2	3	4	5	6	7	8	9	10
heme geometry	ZGK^o	ZGK^o	ZGK^i	ZGK^i	K^i	K^i	K^i	K^i	K^i	K^i
H_2O	no	yes	yes	no	no	no	no	no	no	no
Fe-C distance (Å)	1.84	1.84	1.84	1.64	1.64	1.54	1.54	1.54	1.54	1.40
angle ß of Fe-C-O	180	180	180	180	180	180	165	150	135	135
$\Delta E_Q{}^a$ (mm/sec)	1.85	2.25	1.75	1.15	0.87	0.73	0.66	0.60	0.50	0.60
δ^b (mm/sec)	0.70					0.37			0.30	

a) sign of calculated EFG is positive in agreement with experiment.

b) isomershifts given relative to α-Fe; obtained by using as reference compound $Fe(CN)_6^{-4}$ and as isomershift calibration constant $\alpha = \Delta\delta/\Delta\rho(0)$ the value $-0.35\ a_o^3$ mm/sec.

i) iron in the plane of the four heme-nitrogens.

o) iron out of the plane of the four heme-nitrogens.

We find that the original molecular geometry of Zerner et al. for HbCO (no. 1 in table II) results in a ΔE_Q -value which is far from the experimental one. Adding a water molecule to the system increases ΔE_Q (no. 2). Rotating the water around the z-axis in steps of 20° has no significant effect on ΔE_Q . Having the iron in the plane of the four nitrogens reduces ΔE_Q (no. 3). By taking away the water molecule and reducing the Fe-C distance to 1.64 Å (no. 4) ΔE_Q decreases. Therefore we use for the further calculations molecular geometries K, without water, and iron in the plane of the four nitrogens, all three conditions have the effect of reducing ΔE_Q in the direction of the experimental value. For the new Fe-C distance of 1.54 Å we rotate the CO molecule according to fig. 4 from ß = 180° to 135° in steps of 15° (no 6-9). Our estimate of ΔE_Q and δ indicates that the bent geometry is more favourable than the colinear one. A further decrease of the Fe-C distance seems not to be desirable, because ΔE_Q increases again (no. 10).

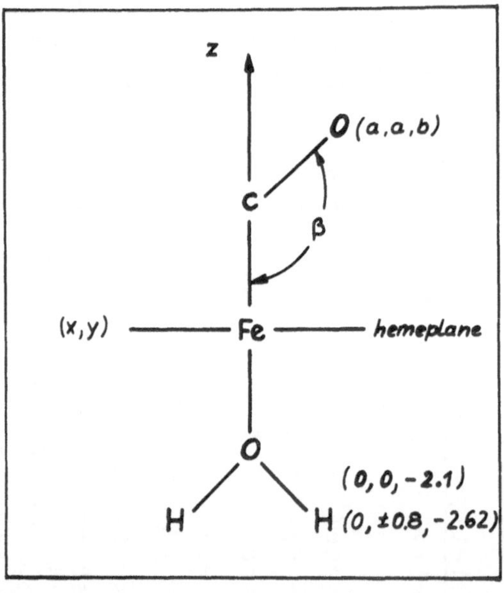

Fig. 4 HbCO - geometry; cartesian coordinates in Å, with a = 1.13 sin (90-ß) sin 45, and b = 1.13 cos (90-ß).

Energies of charge transfer and d-d transitions have been calculated for geometry no. 9 and are found to give reasonable values, contrary to results for geometry no. 1. We further estimate energies for the many electron terms $^5B_2$, ^3E and $^1A_1$ using formulae as obtained from our ligand field calculations /7/. We find for molecular geometry no. 1 the high spin term $^5B_2$ to be the groundstate, and for no. 9 the $^1A_1$ term to be the right groundstate. The latter turns out to be isolated from any excited state by about 2500 cm^{-1}, in agreement with our ligand field results /7,10/.

The results of the present MO and Mössbauer study of HbCO can be summarized:

1. the iron is situated in the heme plane in agreement with recent x-ray studies on the carbonmonoxide complex of erythrocruorine /31/,
2. the bonding between the heme iron and the proximal histidine is weak,
3. the Fe-C distance is somewhat smaller than the typical iron-carbonyl distance of 1.84 Å,
4. the bent geometry of CO with ß = 135° is more favourable than the colinear geometry with ß = 180° . This is in agreement with the x-ray studies on erythrocruorine /31/ and glycera /32/ hemoglobin, for which a Fe-C-O bond angle of 145° ± 15° was found.

<u>3.) HbO_2.</u> For oxyhemoglobin preliminary results are presented. Four different heme-O_2-geometries as shown by fig. 5 have been investigated so far. For all of them the high spin term $^5B_2$ is estimated to be groundstate and to be isolated from excited states by about 1500 cm^{-1}. This is the result of relatively close lying Hückel energy terms above the highest doubly occupied Hückel MO, necessitating the use of configuration interaction. Such CI calculations for HbO_2 are in progress.

For geometries no. 1 and no. 2 a difference in the total energies is found to be about 1500 cm^{-1} indicating that the model of explaining the temperature dependence of the HbO_2 -quadrupole splitting by rotation of O_2 /4,14/ is relatively improbable.

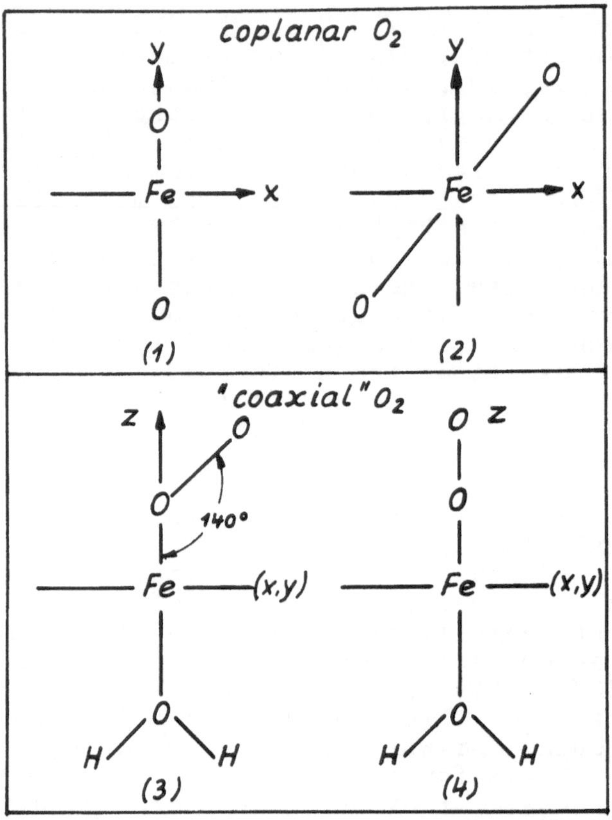

Fig. 5 Molecular geometries used for MO study of HbO_2.
For (3) a minimum inthe total energy was found in accordance with recent IEH results by M. P. Halton Theor. Chim. Acta 23, 208 (1971)

Discussion

From the interpretation of isomer shifts and quadrupole splittings using theoretical approaches such as ligand field LF and molecular orbital MO theory it is possible to derive some knowledge about the electronic state of iron and about the possible geometry of the compound under study.

For carbonmonoxyhemoglobin equal results are obtained from LF and MO theory concerning the energy sequence of electronic states. The MO approach, however, turns out to be much more powerful concerning molecular geometries.

Mössbauer data of deoxyhemoglobin are explained on the basis of LF calculations by taking into account a low lying intermediate spin state. This result may be used to describe a possible pathway for the reaction of Hb with O_2 /11/. O_2 is known to form antiferromagnetically coupled molecule complexes in its triplet ground state $^3\Sigma_g^-$, e.g. $(O_2)_2$ /33/ or (O_2) in solid oxygen /34/. One can expect that antiferromagnetic spin coupling between the O_2 molecule in its $^3\Sigma_g^-$ ground state and the iron of Hb varies the thermal equilibrium between the ^3E triplet and the $^5B_2$ quintet states, resulting in a small increase in the heme plane ligand field, as the iron, pentacoordinated in Hb, approximates hexacoordination during the approach of O_2. It was shown (see formulas 32 a-c of reference /7/) that an increased heme plane ligand field lowers the energetic gap between ^3E and $^5B_2$. Therefore one can suppose that antiferromagnetic spin coupling results in considerable spin quenching:

$$Fe(^5B_2, {}^3E) + (^3\Sigma_g^-)O_2 \rightleftharpoons \underset{S\approx 1(\downarrow\downarrow)}{Fe(^3E, {}^5B_2)} --- \underset{S=1(\uparrow\uparrow)}{(^3\Sigma_g^-)O_2}.$$

This may initiate the oxygenation, however, it seems not to be sufficient to explain the thermodynamically stable state of HbO_2, which is usually described by a double-bond system /35/. Such a double-bond system can be formed for both classical models for the oxygen-hemoglobin bonding (Pauling /13/ and Griffith /12/) by combining iron in the ^3E triplet state with excited triplet states of the O_2 molecule /36/. They can be formed without alteration of the spin multiplicity of the components during the formation or break up of the oxygenated adduct:

$$\underset{S\approx 1(\downarrow\downarrow)}{Fe(^3E, {}^5B_2)} --- \underset{S=1(\uparrow\uparrow)}{(^3\Sigma_g^-)O_2} \rightleftharpoons \underset{S=0}{FeO_2}\quad .$$

This aspect is in agreement with the experimental quadrupole splitting and isomer shift data of HbO_2 (table I) since the ^3E level causes a negative EFG at the iron nucleus, and compounds with ^3E groundstate have comparable experimental isomer shifts /37,38/. It is also in agreement with the proposal that the bond length in transition metal-O_2 -systems are best explained assuming higher triplet states of O_2 /37/.

The conservation of the spin multiplicity may be the crucial point for the easy kinetic reversibility of the oxygenation of hemoglobin.

Acknowledgement

I am very grateful to Dr. H. Eicher with whom together the ligand field calculations were performed at the Technische Universität in Munich, West Germany. I am also very grateful to Prof. Dr. F. E. Harris with whom the molecular orbital calculations were carried out at CECAM, Orsay, France, and at the University of Utah, Salt Lake City, USA, with the Univac 1108 of the CC/UU in Salt Lake City. I am further indebted to Prof. Dr. F. E. Harris and his wife Ms. Grace Harris for their friendly hospitality during my stay in Salt Lake City. Useful discussions with Prof. Dr. U. Gonser and Drs. H. Formanek, D. Klint, Y. Maeda, and D. L. Williamson are acknowledged. I thank Ms. R. Bubel and Ing. J. Welsch for preparing the manuscript.

References

/1/ U. Gonser, R. W. Grant, J. Kregdze
Appl. Phys. Lett. 3, 189 (1963), and
Science 143, 680 (1964)

/2/ W. Karger
Berichte der Bunsengesellschaft für physikalische Chemie 68, 793 (1963)

/3/ U. Gonser, R. W. Grant
Biophys. J. 5, 823 (1965)

/4/ G. Lang, W. Marshall
Proc. Phys. Soc. (London) 87, 3 (1966)

/5/ H. M. Kappler, A. Trautwein, A. Mayer, H. Vogel
Nucl. Instr. Methods 53, 157 (1967)

/6/ Conference on the Physical Properties of Iron

Proteins, J. E. Maling and M. Weissbluth, eds. Stanford University, Stanford, Calif., 1967

/7/ H. Eicher, A. Trautwein J. Chem. Phys. 50, 2540 (1969)

/8/ H. Eicher, A. Trautwein J. Chem. Phys. 52, 932 (1970)

/9/ A. Trautwein, H. Eicher, A. Mayer J. Chem. Phys. 52, 2743 (1970)

/10/ A. Trautwein, H. Eicher, A. Mayer, A. Alfsen, M. Waks, J. Rosa, Y. Beuzard J. Chem. Phys. 53, 963 (1970)

/11/ A. Trautwein, P. Schretzmann Proceedings of the International Conference on Hyperfine Interactions and Excited Nuclei, Goldring and Kalish, eds., p. 881 (1971)

/12/ J. S. Griffith Proc. Roy. Soc. (London) A235, 23 (1956)

/13/ L. Pauling Nature 203, 182 (1964)

/14/ G. Lang Quart. Rev. Biophys. 3, 1 (1970)

/15/ L. Pauling, C. D. Coryell Proc. Nat. Acad. Sci. USA 22, 210 (1936)

/16/ M. Zerner, M. Gouterman, H. Kobayashi Theoret. chim. Acta 6, 363 (1966)

/17/ M. Weissbluth, J. E. Maling J. Chem. Phys. 47, 4166 (1967)

/18/ A. Trautwein, F. E. Harris, Y. Maeda, H. Formanek, to be published

/19/ A. Trautwein, F. E. Harris submitted to J. Chem. Phys.

/20/ A. Trautwein, F. E. Harris, J. R. Regnard, Y. Maeda submitted to Phys. Rev. B

/21/ A. Trautwein, F. E. Harris, submitted to Phys. Rev. B

/22/ C. E. Moore Atomic Energy Levels, Natl. Bur. Std. Circ. No. 467, 2 (1952)

/23/ H. Eicher Z. Physik 169, 178 (1962)

/24/ R. Rein, N. Fukuda, H. Win, G. A. Clarke F. E. Harris J. Chem. Phys. 45, 4743 (1966)
R. Rein, G. A. Clarke, F. E. Harris Quantum Aspects of Heterocyclic Compounds in

Chemistry and Biochemistry, II, Israel Academy of Sciences and Humanities, Jerusalem, 1970
/25/ W. H. Flygare
J. Chem. Phys. 43, 789 (1965)
/26/ R. E. Watson
Solid State and Molecular Theory Group. Technical Report No. 12, MIT, 1959 (unpublished)
/27/ J. Blomquist, B. Roos, M. Sandbom
J. Chem. Phys. 55, 141 (1971
/28/ D. A. Shirley
Rev. Mod. Phys. 36, No. 1, Part 2, 339 (1964)
/29/ E. Clementi
Supplement to IBM J. Res. Dev. 9, 2 (1965)
/30/ D. F. Koenig
Acta Cryst. 18, 663 (1965)
/31/ R. Huber, O. Epp, H. Formanek
J. Mol. Biol. 52, 349 (1970)
/32/ private communication of E.A. Padlan
/33/ S. Arnold, N. Finlayson, E. A. Ogryzlo
J. Chem. Phys. 44, 2529 (1966)
/34/ C. Barrett
J. Chem. Phys. 47, 592 (1967)
/35/ E. Bayer, P. Schretzmann
Structure and Bonding, Vol. 2, p. 181
C. K. Jørgenson ed., Springer-Verlag, Berlin 1967
/36/ G. Herzberg
Spectra of Diatomic Molecules, p. 446, Van Nostrand Co., Princeton, 1967
/37/ B. W. Dale, R. J. P. Williams, P. R. Edwards, C. E. Johnson
J. Chem. Phys. 49, 3445 (1968)
/38/ E. König, K. Madeja
Inorganic Chemistry 7, 1849 (1968)
/39/ R. Mason
Nature 217, 543 (1968)

THE STUDY OF INERT-GAS MATRIX ISOLATED ATOMS

P. H. Barrett and H. Micklitz*

Department of Physics

University of California, Santa Barbara, Calif.

INTRODUCTION

The isolation of atoms in a solid rare-gas matrix has been a useful technique in optical and EPR spectroscopy. These experiments have shown that the weak binding of the matrix does not change the electron configuration from that of the free atom. Only small changes in atomic energy levels and in the hyperfine interaction occur. In 1961 Jaccarino and Wertheim[1] proposed that this technique be used to measure the Mössbauer effect isomer shift of the "almost free" Fe-57 atom with the atomic configuration $3d^6 4s^2$. It was many years later before the first successful rare-gas-matrix-isolation (RGMI) Mössbauer experiment was reported.[2] In this experiment iodine molecules (I_2) were imbedded in a solid argon matrix. We have followed up on the suggestion of Jaccarino and Wertheim and have used the RGMI/Mössbauer technique to study Fe-57 and Sn-119, in some cases with a variety of gasses (Ne, Ar, Kr, Xe, and N_2) and over a range of temperatures (1.4K to 27K). These experiments and other recent RGMI/Mössbauer experiments are discussed in the following sections.

EXPERIMENTAL TECHNIQUE

The procedure we have used is to condense simultaneously the inert-gas atoms and the atoms under study onto a cold (usually 4.2K) surface and then use this sample as a source or absorber with the usual Mössbauer spectroscopy apparatus. It is the experimental details of this simultaneous condensation that determine the success or failure of the experiment. A detailed description of the apparatus and the operating procedures is given in Ref. 3. We have found a clean cryostat and a high-speed vacuum pumping system essential to the production of good samples. With a gas deposition rate of about ten atomic layers per second and a vapor pressure of the metal sample inside the furnace of about 10^{-3} torr, an adequate sample of about one atomic percent of metal atoms is produced in several hours. The absorption of the gamma ray in the gas matrix is usually the limiting factor on the thickness of the layer. By measuring the absorption of the 6-keV iron X ray in the sample we continuously monitor the deposition of the gas. At a residual gas pressure of 10^{-7} torr in the cryostat, we estimate the oxygen concentration in the samples to be at least an order of magnitude down from the metal atom concentration.

ISOMER SHIFT

The Mössbauer spectra of iron[3-5] and tin[6] contain a line whose area is proportional to the first power of concentration, this we identify with an isolated atom (monomer) and a pair of lines whose area is proportional to the square of the concentration, these we identify with the quadrupole interaction of a nearest-neighbor metal atom (dimer). The dimer results will be discussed later.

The Mössbauer-effect isomer shift is given by Shirley[7] as $\delta=(4/5)\pi Ze^2R^2cE_\gamma^{-1}S'(Z)[\Psi_A^2(0)-\Psi_S^2(0)](\Delta R/R)$ where ΔR is the change in charge radius R between the excited and ground state of the nucleus. $\Psi^2(0)$ is the nonrelativistic electron density at the nucleus and the subscripts refer to the source (S) and absorber (A). $S'(Z)$ is a term that corrects the electronic wave functions for relativistic effects. The use of this equation to relate isomer shifts to electron densities depends on an accurate value of $\Delta R/R$, but, unfortunately, the values obtained from experiments with solid compounds differ by as much as a factor of three. In these experiments there is an uncertainty as to the applicability of the free-atom wave functions and to the choice of electron configurations for particular reference compounds. The use of the RGMI technique reduces if not eliminates these uncertainties.

Figure 1 shows a plot of the calculated relativistic electron densities for iron [$\Psi^2(0)=S'(Z)\Psi^2(0)$, where we have used $S'(Z)=1.29$[7] and the values of $\Psi^2(0)$ given by Blomquist et al[8]] vs isomer shift of Fe-57. The isomer shift of the rare-gas matrix isolated Fe-57 atom (monomer) is plotted with the free-atom value of $\Psi^2(0)$ for the $3d^6 4s^2$ electron configuration. Optical absorption experiments[9,10] with RGMI iron atoms have shown that the ground state configuration is $3d^6 4s^2$ and that the matrix perturbations are very small. We have made overlap calculations that show that the maximum increase in the 4s electron density at the iron nucleus caused by the neighboring rare-gas atoms is five percent or less. This agrees with the observation[11] of a 4 to 5 percent change in the ESR hf coupling constant A of $Cu(3d^{10}4s^1)$ when placed in a rare-gas matrix. The second important point in Fig. 1 is obtained from a Mössbauer experiment in which the Co-57 source atoms are imbedded in a xenon matrix.[12] In addition to a line with an isomer shift indicating that the 14-keV gamma ray decay occurred in an iron atom with a $3d^6 4s^2$ configuration, a second line of al-

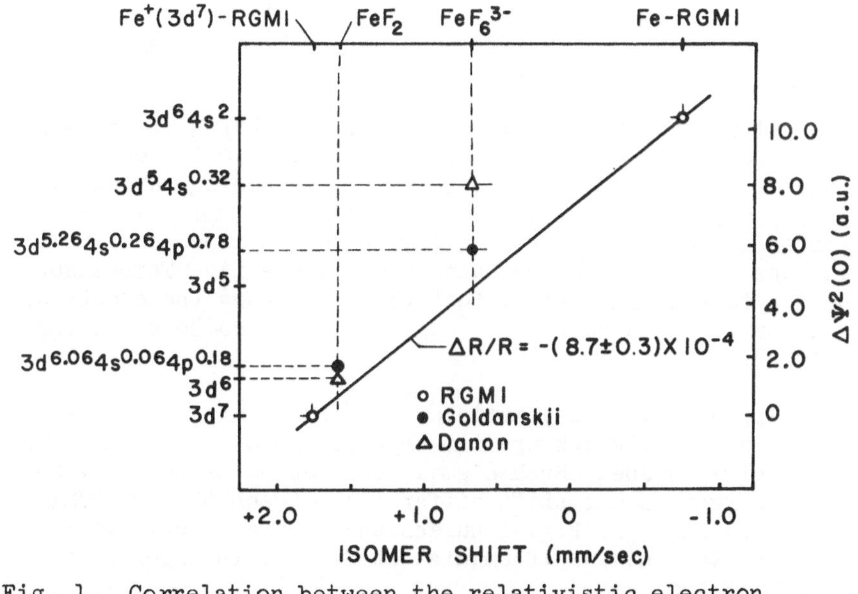

Fig. 1. Correlation between the relativistic electron density $\Psi^2(0)$ at the nucleus and the isomer shift for Fe-57. Isomer shift values are given relative to iron metal at 300K.

most equal intensity appeared with an isomer shift different from the first line by $\Delta\delta(Fe-Fe^+)=2.53\pm0.08$ mm/sec. We have developed an argument that identifies this second line with the gamma decay occurring in the $Fe^+(3d^7)$ ion. This ion could be produced as an Auger-electron aftereffect following the nuclear electron capture decay of Co-57. The $3d^7$ configuration has the lowest value of $\Psi^2(0)$ as calculated by Blomquist et al[8] of all the likely higher charged states produced by Auger transitions. These calculations give for the difference in the non-relativistic total electron density at the iron nucleus between the atomic configuration $Fe(3d^6 4s^2)$ and $Fe^+(3d^7)$, the value of $\Delta\Psi^2(0)=8.0$ a.u. This gives $\Delta R/R = -(8.7\pm0.3)\times10^{-4}$ for the 14.4-keV γ transition in Fe-57. (The error includes only the experimental error.) Taking into account the small effect of the rare-gas matrix on the 4s electron density would decrease this value of $\Delta R/R$ by about 5%.

Although we are quite confident in the identification of the monomer isomer shift with the $\Psi^2(0)$ of the $3d^6 4s^2$ electron configuration (with less than 5% corrections for matrix effects) we can not exclude decay aftereffects in the lattice as the origin of the source produced state. Further work (e.g. looking for the delayed transition from the $3d^7$ state with optical spectroscopy) is needed to fully understand the decay aftereffects in rare-gas matrices.

Also shown in Fig. 1 are the isomer shifts of the most ionic ferrous compound (FeF_2) and the most ionic ferric complex (FeF_6^{3-}) together with the atomic configurations as proposed by Danon[13] and Goldanskii.[14] Our RGMI calibration points suggest that $\Psi^2(0)$ in the above mentioned compounds is less than proposed by Danon and Goldanskii. This might be due to a solid state effect that decreases the electron density in these compounds (e.g. an increased 3d shielding effect).

There are several articles[15-18] concerning the $\Delta R/R$ value for Fe-57 which predict $\Delta R/R$ values of about half the above given value. Such a small $\Delta R/R$ value cannot explain our observed isomer shift difference between the two RGMI Fe states shown in Fig. 1 unless the Hartree-Fock calculations by Watson[19] and Blomquist et al[8] are in error by a factor of two.

We have done similar experiments with RGMI Sn-119.[6]

The correlation between the relativistic electron densities at the tin nucleus[20] and the isomer shift of the tin RGMI monomer (atomic configuration $4d^{10}5s^{2}5p^{2}$) and several tin compounds[21] is shown in Fig. 2. Also shown are two points from experiments[22] in which the relative intensities of conversion electrons from the N_I and O shell are measured for SnO_2 and β- tin. Together with the calculated 4s electron density, they obtain in this way the 5s electron density in these compounds.

Fig. 2 shows the uncertainty that still exists in the ΔR/R value for Sn-119. The ΔR/R value of 7.3×10^{-5} might be

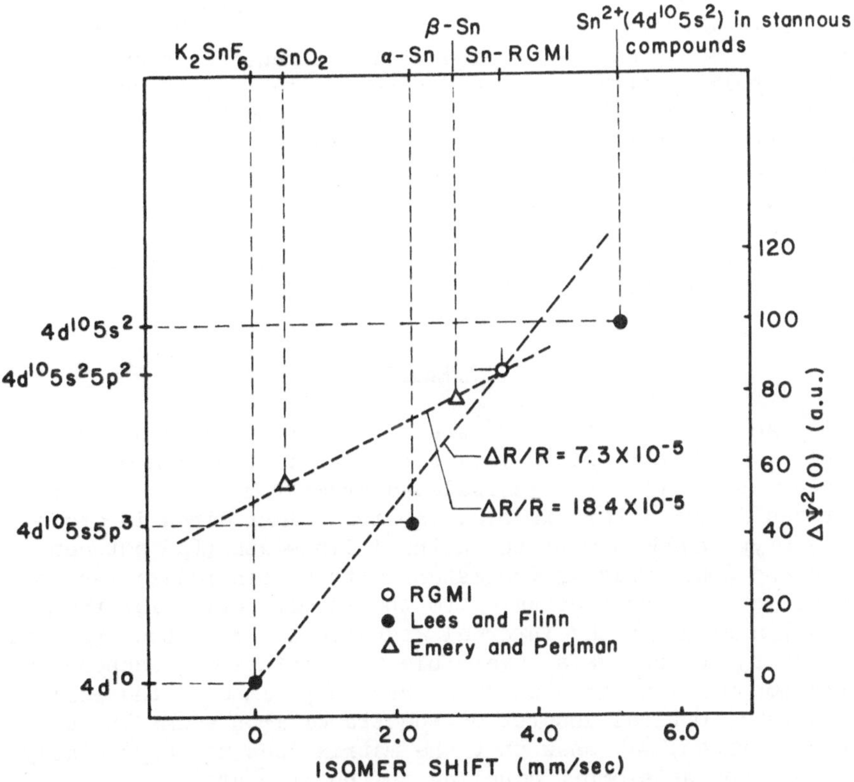

Fig. 2. Correlation between the relativistic electron densities $\Psi^2(0)$ at the nucleus and the isomer shift for Sn-119. Isomer shift values are given relative to K_2SnF_6 at 4K.

a lower limit since it assumes that K_2SnF_6 is completely ionic.

A second RGMI calibration point for Sn-119 seems desirable. However, until now we have not obtained such a second Sn calibration point from a RGMI Sn-119m source experiment.

DIMERS

When impurity atoms are randomly distributed on substitutional sites in a cubic rare-gas matrix, there is an appreciable probability (about 20 percent for 2 percent atomic concentration) that an impurity atom will have another impurity atom as a nearest neighbor. This pair of impurity atoms (dimer) may or may not be a bound state; however, the axial symmetry may remove the atomic degeneracies and produce quadrupole splitting of the nuclear states. This splitting has been observed in both RGMI Fe-57 [ΔE_Q=4.05±0.04 mm/sec] and Sn-119 [ΔE_Q=3.5±0.25 mm/sec]. Free atom wave functions have been used to calculate the electric field gradients at the nucleus and, within the accuracy of this assumption, we have obtained the quadrupole moment for the $I = 3/2$ excited states of Fe-57 and Sn-119 the values of $Q = \pm 0.213$ b and $Q = \pm 0.065$ b respectively.

LINEWIDTH

An analysis of the Mössbauer spectra linewidths as a function of temperature yields further information on the interactions between the isolated atom or molecule and the frozen matrix. For example, the iron dimer lines in argon and krypton have almost the natural linewidth (Γ_0) between 1.4 and 20K. This is consistent with a spin relaxation time that is very short compared to the Larmor period for the Fe-57 nucleus in the internal magnetic field of the iron atom. However, we observe a reversible temperature dependence for the monomer linewidth which we have explained in the following way: Optical absorption spectra of iron atoms in rare-gas matrices[9,10] show that the matrix does not appreciably change the spin-orbit coupling in the iron atom and, therefore, the orbital momentum is not quenched, and J is a good quantum number. From this it follows that the hyperfine magnetic field H_{hf} at the iron nucleus in a rare-gas matrix is approximately that of the free iron atom. This has been determined by an atomic beam NMR experiment[23] to be $H_{hf}=(1.11\pm0.02)\times10^6$ Oe. The large hyperfine splitting (splitting of the two outer lines $\Delta\approx35$ mm/sec) that would be expected from such a field is not observed. To explain this, we assume a spin-lattice relaxation time $T_1 \ll 1/\omega_L$ (ω_L is the nuclear Larmor frequency in H_{hf}). In this region the normal six-line pattern collapses to a single line with a half-width of $\Gamma=\Gamma_0+\Delta\Gamma$, where $\Delta\Gamma=(\omega_L/2\pi\,T_1^{-1})\Delta$.[24] From the measured 1/T dependence of $\Delta\Gamma$, it follows that $T_1^{-1}\propto T$, which shows that the direct phonon process is dominant for the spin relaxation of Fe-57 in rare-gas matrices. For $H_{hf}=1.1\times10^6$ Oe and $1/T_1=AT$, we find that $AT=2.5\times10^9$T sec^{-1} for krypton and xenon and $AT=1.8\times10^9$T sec^{-1} for argon above 4K. Below 4K in argon, the data are consistent with $\Delta\Gamma\propto\tanh(\Delta E/2kT)$, which is expected for the direct phonon spin-relaxation when $\Delta E\sim kT$ (ΔE is the crystal-field splitting between the atomic levels which are involved in the spin-lattice relaxation process).

MÖSSBAUER TEMPERATURE

In addition to spin-lattice relaxation effects, there are other lattice phenomena that can be explored by the RGMI technique. The temperature dependence of the Mössbauer f factor allows the calculation of a characteristic Debye temperature (Mössbauer temperature Θ_M). For the iron monomer we find Θ_M about 60K for argon, krypton, and xenon. This value is lower than expected from correcting the specific heat Debye temperatures of the rare-gas matrices for the difference in mass between the host lattice atom and the impurity atom.

MIGRATION EFFECTS

In addition to the longer spin-lattice relaxation time noted above for argon, we find that the dimer to monomer intensity ratio is for Fe-57 three times and for Sn-119 1.5 to 2 times the value expected from random distribution of impurity atoms (krypton and xenon give the expected dimer/monomer ratio). This can be explained by the surface migration of impurity atoms during the deposition of the argon matrix. There is probably no bulk diffusion of impurity atoms as we see no increase in the iron dimer relative to the iron monomer when the argon matrix is heated up to 20.5K after deposition.

Ne AND N_2 MATRIX

The low Debye temperatures of rare-gas solids limit the Mössbauer absorption experiments to γ transitions with low γ energy. The relatively small absorption lengths for the low energy gamma rays limit the thickness of the rare-gas matrix. Therefore, from the standpoint of small gamma ray absorption, the lightest rare-gas (neon) would seem desirable as an isolating matrix. We have found, as have others,[11,25] that atoms trapped in a neon matrix give confusing results. For Fe-57 in a neon matrix, we do not observe the monomer but the dimer is clearly identifiable. Also, there appears what is probably the result of hyperfine interactions. This suggests that the surface migration of impurity atoms which was proposed above as occurring in argon becomes very rapid in neon and that iron atoms almost always combine with other iron atoms or impurities such as oxygen in the neon matrix.

Nitrogen is another gas matrix that has a long absorption length for γ rays. It has been used to study trapped atoms by optical spectroscopy and the results show that the matrix has only a small effect on the atomic configuration. We have studied Fe-57 and Sn-119 in solid nitrogen.[26] Both these impurity atoms show the same isomer shifts in solid nitrogen as in the RGMI experiments. However, they both experience a quadrupole splitting which we propose is due to either the non-cubic point symmetry of the nitrogen lattice or the distortion of the nitrogen lattice by the impurity atom.

HIGH CONCENTRATIONS

The experiments mentioned above have been done in the impurity atomic concentration range from 0.3 to 3%. The attempt to increase the Sn-119 concentration produced an absorption spectrum with an isomer shift close to that of β-tin. Similar attempts at higher concentrations of Fe-57 produced a six-line spectrum with slightly less splitting than bulk iron. These effects at higher concentration may be the result of clustering of impurity atoms or it may indicate the onset of long range ordering in a random distribution.

FURTHER EXPERIMENTS

Other experiments with the matrix isolation technique that look promising are (1) using molecules containing Fe-57 and Sn-119 and (2) trapping other Mössbauer nuclei. McNab et al[27] have studied $FeCl_2$ in argon and xenon and have observed a small (0.62 mm/sec) quadrupole splitting of Fe-57 in the isolated molecule. In the study of molecules it will be necessary to establish just which molecular species are evaporating from the furnace. This identification can probably be done with optical or infrared absorption spectroscopy and indeed much work has already been done with trapped molecules with these techniques. The low Mössbauer temperature of the RGMI atoms puts a limitation on the gamma ray energy or more precisely the nuclear recoil energy. There are thirteen Mössbauer nuclei with recoil energies less than 7×10^{-3} eV (i.e., $E_R \lesssim 2k\Theta_D$). Four of these (Kr-83,[28] I-129, Fe-57, and Sn-119) have been isolated in a rare-gas matrix and the next most likely candidates are Eu-151, Dy-161, and Sm-149.

The RGMI technique offers to be a useful tool when used with other spectroscopic methods in the study of atomic, molecular, and lattice phenomena.

REFERENCES

*On leave from Department of Physics, Technische Universität München, Munich, Germany.

1. V. Jaccarino and G.K. Wertheim, in The Mössbauer Effect, edited by D.M.J. Compton and A.H. Schoen (Wiley, New York, 1962), p.260.

2. S. Bukshpan, C. Goldstein and T. Sonnino, J. Chem. Phys. 49, 5477 (1968).

3. T.K. McNab and P.H. Barrett, in Mössbauer Effect Methodology (Plenum, New York 1971), Vol. 7, p. 59.

4. P.H. Barrett and T.K. McNab, Phys. Rev. Letters 25, 1601 (1970).

5. T.K. McNab, H. Micklitz and P.H. Barrett, Phys. Rev. B4, 3787 (1971).

6. H. Micklitz and P.H. Barrett, Phys. Rev. B5, 1704 (1972).

7. D.A. Shirley, Rev. Mod. Phys. 36, 339 (1964).

8. J. Blomquist, B. Roos and M. Sundborn, J. Chem. Phys. 55, 141 (1971).
9. D.M.Mann and H.P.Broida, J. Chem. Phys. 55, 84 (1971).
10. H.Micklitz and P.H.Barrett, Phys. Rev. B4, 3845 (1971).
11. P.H.Kasai and D.McLeod, Jr., J. Chem Phys. 55, 1566 (1971).
12. H.Micklitz and P.H.Barrett, Phys. Rev. Letters, 28, 1547 (1972).
13. J. Danon, in Chemical Applications of Mössbauer Spectroscopy, edited by V.I.Goldanskii and R.H.Herber (Academic, New York 1968), p.173.
14. V.I.Goldanskii, E.F.Makarov and R.A.Stukan, Teor. Eksp. Khimiga 2, 504 (1966) Translation: Theor. Exp. Chem. 2, 382 (1966).
15. E.Simanek and A.Y.C.Wong, Phys. Rev. 166, 348 (1968).
16. F.Pleitner and B.Kolk, Phys. Letters 34B, 296 (1971).
17. R.Ruegsegger and W.Kündig, Phys. Letters 39B, 620 (1972).
18. R.R.Sharma and A.K.Sharma, Phys. Rev. Lett. 29, 122 (1972).
19. R.E.Watson, MIT Solid State and Molecular Theory Group Technical Report No. 12 (unpublished); Phys. Rev. 119, 1934 (1960).
20. D. Liberman, J.T.Waber and D.T.Cromer, Phys. Rev. 137, A27 (1965), J.T.Waber (private communication to J.K.Lees and P.A.Flinn).
21. J.K.Lees and P.A.Flinn, J. Chem. Phys. 48, 882 (1968).
22. G.T.Emery and M.L.Perlman, Phys. Rev. B1, 3885 (1970).
23. L.S.Goodmann and W.J.Childs, Bull. Am. Phys. Soc. 9, 12 (1964).
24. G.K.Wertheim, Phys. Rev. 121, 63 (1961).
25. W.Keune and E.Lüscher, Ann. Univ. Saraviensis, Sci. 8, 92 (1970).
26. H.Micklitz and P.H.Barrett, Appl.Phys.Lett. 20, 387 (1972).
27. T.K.McNab, D.H.W.Carstens, D.M.Gruen and R.L.McBeth, Chem. Phys. Letters 13, 600 (1972).
28. S. Ruby and H. Selig, ANL-7108, p.17; M.Pasternak, A. Simopoulos, S.Bukshpan and T.Sonnino, Phys. Letters 22, 52 (1966).

A STUDY OF THE MÖSSBAUER EFFECT FOR GOLD*

Louis D. Roberts and J. F. Prince
University of North Carolina, Chapel Hill

D. J. Erickson**
University of North Carolina, Chapel Hill and
Los Alamos Scientific Laboratory

In the study of metals and alloys through the use of the Mossbauer effect, gold is of particular interest, for it is the only Mössbauer element which is a noble metal. It would be naive to suppose that any metallic system is simple. Yet, among the transition elements, the noble metals, pure or in alloys, may be the ones that can be discussed in terms of available theory with greatest reliability.

In this article, we will review some of the gold Mössbauer work with which we and our colleagues have been concerned, and indicate how this work relates to the screening of atoms in alloys. We will not give a review of the literature for the Mössbauer effect for Au here, but extensive references are given in the papers to which we will refer. In the first part of the paper, we will speak primarily of the isomer shift, ΔE, and in the later part, of the recoilless fraction, $f(T)$.

The Isomer Shift

(a) The ^{197}Au Isomer Shift for Au Alloyed in Ag, Cu, Pd, and Pt.

In early gold Mossbauer work,[1,2] it was found that the isomer shift ΔE for Au in a dilute solid solution in Ag was $+2.14 \pm 0.05$ mm/sec. This is relative to pure Au, for a moving source and a stationary alloy absorber. This ΔE may be compared with a recent measurement[3] of the Mössbauer line width $2\Gamma_0 = 1.846 \pm 0.012$ mm/sec.

The isomer shift[4] is described in good approximation by

$$\Delta E = G(|\psi_{alloy}(0)|^2/|\psi_{Au}(0)|^2 - 1) \qquad (1a)$$

with

$$G = B|\psi_{Au}(0)|^2(\langle r_e^2\rangle - \langle r_g^2\rangle) \, . \qquad (1b)$$

Here $|\psi_{Au}(0)|^2$ describes the valence band electron charge density at a Au nucleus in gold, $|\psi_{Alloy}(0)|^2$ describes this charge density for the gold nuclei in an alloy. $\langle r_e^2\rangle$ and $\langle r_g^2\rangle$ are the average squared radii of a ^{197}Au nucleus in its excited and ground states and B is a constant which may be calculated.

At the time of the first measurements of ΔE for Ag(Au) it was not possible to say whether $|\psi_{Au}(0)|^2 > |\psi_{alloy}(0)|^2$ and $\langle r_g^2\rangle > \langle r_e^2\rangle$ or just the opposite for both terms. One of these alternatives was true, however, because experimentally, $\Delta E > 0$. A choice between these alternatives was made through a study of screening in the alloys.[2]

Many extensive theoretical calculations have been made over a period of many years of the screening of a charge by an electron gas, but comparitively few experimental studies have been made of the screening charge distribution itself.[5] It appeared that the isomer shift for Au in an alloy should be interpreted as an aspect of the screening of the gold. Thus a study of ΔE should provide a way of investigating the screening charge.

Screening concepts had been used to calculate the electrical resistence due to the solute in dilute alloys.[6] We thus sought to calculate the isomer shift ΔE for say a Au impurity in Ag and to correlate ΔE with $\Delta\rho/c$, the impurity electrical resistivity per atomic percent.[2] $\Delta\rho/c$ is proportional to the transport cross section σ_t of the impurity for conduction electrons; and σ_t for say Au in Ag may be calculated readily from a measured value of $\Delta\rho/c$ for the alloy.

In more pictorial terms, if a gold impurity atom, when placed in a metallic host, produced a disturbance or shift of electrical charge in the gold atom and around it in the conduction band of the host, then both the isomer shift and the impurity electrical resistivity should be aspects of this shift of electric charge. It should be possible to demonstrate a correlation between the isomer shift (the

γ-ray energy) and the electrical resistivity if a suitable potential for Au in, for example, Ag could be found. A suitable potential was found in the following way.

By methods similar to those of Friedel,[6] Blandin,[7] and Daniel,[8] the measured $\Delta\rho/c$ (or σ_t) and the Friedel sum rule were used together to estimate parameters for a model potential for an atom of Au in say a Ag host.

The Friedel sum rule[6] is given by

$$\Delta Z = \frac{2}{\pi} \sum_{\ell}(2\ell + 1)\delta_{\ell} \qquad (2)$$

and

$$\sigma_t = \frac{4\pi}{k_F^2} \sum_{\ell} (\ell + 1)\sin^2(\delta_{\ell} - \delta_{\ell+1}). \qquad (3)$$

Here ℓ is the angular momentum quantum number of a conduction electron partial wave, δ_{ℓ} is the phase shift due to scattering of the ℓth partial wave at the Fermi energy by the Au impurity, ΔZ is the valence difference between the impurity and the host atoms, and k is a wave vector of the conduction electrons. k_F is the value of this vector at the Fermi energy.

σ_t for Au in Ag had been measured. Since it was not equal to zero, some of the δ_{ℓ} were also not equal to zero. The valences of Au and Ag, however, were each equal to one and ΔZ thus was equal to zero. Thus from Eqs. 2 and 3, to make $\Delta Z = 0$ with $|\delta_{\ell}| \neq 0$, it was clear that some of the δ_{ℓ} must be positive and some negative.

To obtain a model potential which could give δ_{ℓ} of either sign for different ℓ, we postulated a model potential W of operator form[2]

$$W = \sum_{\ell} v_{\ell,R}|\ell\rangle\langle\ell| \ . \qquad (4)$$

In the application of this potential to the calculation of the isomer shift, ΔE, the $v_{\ell,R}$ were taken to be square wells of radius R, the Wigner-Seitz radius for Au, and $|\ell\rangle\langle\ell|$ is a projection operator for the ℓth partial wave. Thus a given v_{ℓ} can scatter only the ℓth partial wave. The well depth for each v_{ℓ} then has a direct correspondence with each phase shift, δ_{ℓ}. δ_{ℓ} will be positive if $v_{\ell,R}$ is negative and vice versa. V. Heine and coworkers[9] have used a model potential of similar form.

The above ideas were used in the calculation of the isomer shift as follows: It was assumed that only δ_0 and δ_1 were finite, that all $\delta_\ell = 0$ for $\ell \geq 2$. Then, using the measured σ_t, Eqs. 2 and 3 gave us two equations in the two unknowns, δ_0 and δ_1, the s and p wave phase shifts for the conduction electrons at the Fermi surface, E_F. We solved these equations in particular for δ_0. Then, from Eq. 4 we calculated the v_0 which would produce this δ_0 at E_F. It was found[2] that $v_0 = -1.1$ eV for Ag(Au).

The charge density at the gold nucleus, $|\psi_{alloy}(0)|^2$ is due only to the $\ell = 0$ partial wave, i.e., only v_0 of Eq. 4 would contribute to ΔE. It was assumed then that this $v_0 = -1.1$ eV, obtained from δ_0 at E_F, was independent of electron energy within the valence band of the Ag host. v_0 has since been shown to be approximately independent of electron energy,[10] (constant within about 10%) through the energy range, $0 \leq E \leq E_F$. This v_0 was used to calculate $|\psi_k(0)|^2$, the charge density at the gold nucleus for each state k throughout the band. $|\psi_k(0)|^2$ was then integrated over the valence band, from 0 to k_F, of the Ag host to obtain a value for the $|\psi_{alloy}(0)|^2$ of Eq. 1.

In this way we obtained a theoretical estimate of the charge density at the Au nucleus from (i) the Friedel sum rule together with (ii) an electrical resistance measurement. The calculation was more detailed and complete[2] than our brief discussion here can indicate.

We followed the above procedure for dilute solid solutions of Au alloyed in Ag, Cu, Pd and Pt hosts. We found that all of the isomer shift data could be fitted together quite well[2] with $G = +(8 \pm 1)$ mm/sec., Eqs. 1. For this case $\langle r_e^2 \rangle > \langle r_g^2 \rangle$.

The form of the operator potential W, Eq. 4, is perhaps the central element in the above correlation of ΔE with $\Delta\rho/c$, and the principal experimental result of this work was the sign and magnitude of G. It was felt that the potential W was basically reasonable and that it would be of value to obtain a check of this model potential. An indirect check of W could be obtained by measuring G by an experimental method which would be quite independent of the above Au alloy study. The method used[11] was a measurement of the pressure dependence of the gold isomer shift.

(b) Measurement of the pressure dependence of the gold isomer shift.

When a pressure of say 75,000 atm. is applied to gold, the density of the metal will increase by about five percent. This fact and the above value for G suggested that ΔE for gold should be measureable as a function of pressure.[11]

One would not necessarily expect $|\psi(0)|^2$ to be proportional to the average metallic density, i.e. to V^{-1} where V is the atomic volume. Because of this, a calculation of the charge density at the gold nucleus as a function of atomic volume was made.[12] This was done in a Dirac, Hartree, Wigner, Seitz approximation, and it was found that

$$|\psi_P(0)|^2/|\psi_{Au}(0)|^2 = [V(0)/V(P)]^\gamma \ . \qquad (5)$$

Here $|\psi_P(0)|^2$ is the charge density at the nucleus for a sample of Au under a pressure P. V(0) and V(P) are the atomic volumes of Au at zero pressure and under an applied pressure P. γ is a constant which was calculated in the Dirac, Hartree, Wigner, Seitz model to have a value of 0.86. Then the pressure, (or volume) dependence of the isomer shift of Eqs. 1 would be given[11,12] by

$$\Delta E \ = \ G[(V(0)/V(P))^\gamma - 1] \qquad (6)$$

The purpose of the high pressure work was thus to measure ΔE(P) as a function of P (or V(P)) and obtain a value of G in this way. If the correlation,[2] of the electrical resistence with ΔE described above in section (a) were correct, then the value of G = 8 ± 1 mm/sec from this alloy isomer shift electrical resistance correlation work and the value of G to be obtained from the high pressure work should agree.

When the high pressure-isomer shift measurements were done, a value for G of + (8 ± 1) mm/sec. was in fact obtained, in agreement with the above alloy study.[11] Thus the model potential W of Eq. 4 was found to correlate $\Delta\rho/c$ with ΔE and also to give a result in agreement with the high pressure work.

The values for G obtained from the alloy studies and the high pressure work depend to some degree on the fact that a Wigner Seitz approximation[12] was used in the interpretation of the measurements in both cases.[2,11] We believe the agreement of the G values from the two kinds of measurements to be significant, and to support the form of the model potential, Eq. 4. However, the value for G may change somewhat when a more complete band theory calculation is used to interpret these alloy and high pressure measurements.[10]

In the above studies, some understanding was gained of the charge density $|\psi_{alloy}(0)|^2$ at gold nuclei in a dilute alloy. It is also of value to study the distribution of charge external to a scattering center.

In a dilute gold alloy ΔE for Au will correspond only to the s-wave phase shift and only to v_0 of Eq. 4. However, in the region around an impurity, $|\psi(r)|^2$ will also in general have a contribution from p, d, and possibly higher partial waves. A study of this subject for Au has been made through measurements of the isomer shift as a function of composition and short range order for the Cu-Au alloy system.[13]

(c) Dependence of the Au isomer shift on short range order in CuAu.

In the above resistance ΔE correlation, section (a), studies were made on dilute alloys with the measured ΔE extrapolated to infinite dilution. The isomer shift was due to charge drawn into a spatially isolated Au atom by the potential, Eq. 4.

In a more concentrated Cu-Au alloy it is convenient to adopt an equivalent but perhaps apparently different view that one is dealing with a sample of gold where some of the gold atoms have been replaced by Cu atoms.[13] In this picture, the Cu is, in a sense, the impurity and the Friedel oscillations[6] due to scattering at the Cu sites will contribute to the charge density at the ^{197}Au nuclei, approximately in an additive way.

When we alter the short range order of Cu and Au atoms around a given Au atom, in an approximate sense, we alter the arrangement of these Friedel oscillations. Thus we change the charge density and ΔE for Au with the short range order of the alloy.

The mathematical statement of this model has been formulated[13] and an equation for the dependence of ΔE on alloy volume, composition, and short range order is given below by Eq. 7.

$$\Delta E = G\{(V_{Au}/V)^{\gamma}[1 + \sum_i m_{Cu}c_i(1 - \alpha_i)\rho_{Cu}(r_i)] - 1\} \quad (7)$$

Here V is the average atomic volume of the alloy and V_{Au} is the atomic volume of pure gold. As in Eqs. 5 and 6, (V_{Au}/V) gives the dependence of the charge density at the gold nuclei on alloy volume. m_{Cu} is the atomic fraction of Cu in the alloy, c_i is the total number of atoms in an ith shell of neighbors around a gold atom at a distance r_i, and the α_i are the short range order parameters. The quantity $\rho_{Cu}(r_i)$ basically describes the amplitude of the charge density oscillations around a Cu atom in the alloy in the ith neighbor shell at a distance r_i from a Au atom.

We have measured[13] ΔE for 23 CuAu alloys over the complete composition range and for these alloys ordered or disordered.

Eq. 7 was found to fit[13] the isomer shift data for the 23 alloys in the ordered phase to better than one percent of the maximum isomer shift, and to fit our ΔE data for those disordered alloys where short range order parameter information was available to within a few percent. In this fit for the 23 alloys, three parameters for a Taylor expansion of the Friedel oscillation function, $\rho_{Cu}(r_i)$, were determined which described $\rho_{Cu}(r_i)$ out to the 2nd neighbor shell. There were no other adjustable parameters in our use of Eq. 7 to describe the data.

For a random CuAu alloy, $(\alpha_i = 0)$, we have the approximate result[13]

$$\sum_i m_{Cu}c_i\rho_{Cu}(r_i) = m_{Cu}[0.186 + (r_1 - \tfrac{a}{\sqrt{2}})\, 0.2554] \quad (8)$$

Here r_1 is the distance to the nearest neighbor shell in the alloy, and $a/\sqrt{2}$ is this distance in pure gold.

The possibility of calculating $\rho_{Cu}(r_i)$ from Eq. 4 is being investigated.

An application of this result has been made to the study of AuNi and AuCuNi alloys which we will sketch in the next section.[14] In this application of Eq. 8 to the AuNi or CuNiAu alloys, we replace m_{Cu} in Eq. 8 by $m_{Cu} + m_{Ni}$.

(d) A Mossbauer study of Au_xNi_{1-x} and $Cu_{x-0.01}Ni_{1-x}Au_{0.01}$ alloys.

Extensive studies of alloys of nickel with the noble metals, copper, silver, and gold, have played a valuable historical and heuristic role in the development of concepts of alloy electronic structure.[15,16] In many of these studies the measurements have been related to alloy magnetic properties and to the often apparently simple dependence of these properties on noble metal concentration. When appropriate, these observations have been interpreted in terms of alloy electronic states of predominantly d-character, and in terms of the dependence on noble metal concentration of the filling of these states.[17] In spite of these numerous studies of noble metal-nickel alloys, however, there is relatively little information about the behavior of the noble metal in these materials. Mossbauer effect studies for Au in Au_xNi_{1-x} and $Cu_{x-0.01}Ni_{1-x}Au_{0.01}$ can give information about the behavior of the noble metal gold in these alloys.

We have measured[14] H_{eff} and ΔE as functions of x. For Au_xNi_{1-x} for $0.01 \leq x \leq 0.2$, and for $0.01 \leq x \leq 0.3$ for $Cu_{x-0.01}Ni_{1-x}Au_{0.01}$, H_{eff} was found to be proportional to the alloy magnetization.

For Au_xNi_{1-x}, ΔE was found to be nearly a linear function of composition whereas ΔE for $Cu_{x-0.01}Ni_{1-x}Au_{0.01}$ showed some structure. In the following, we will discuss ΔE measured for the latter alloy briefly, relating it to Eqs. 7 and 8. The data are shown on the next page in Fig.1.

The sign of ΔE for Au in AuNi or Cu NiAu is of interest. In an early model Stoner[15] and Wohlfarth[16] suggested a kind of electron transfer model for alloys like CuNi or AuNi, and this model has persisted in the literature. In this model, the Ni magnetism was "quenched" to some degree in the presence of the noble metal by the transfer of valence electron charge from the noble metal to the unfilled Ni d-shell. This would lead to a positive charge for the gold in AuNi or CuNiAu. Our result[1,2,11] showed that just the opposite is true. Near the ^{197}Au nucleus, at least, the gold becomes negatively charged in Ni alloys. Further the Au isomer shift shows no inflection near $x \sim 0.6$ where the Ni d-shell becomes effectively full. A comparison of these results, Fig. 1, with Eqs. 7 and 8 is helpful in understanding this behavior of $|\psi_{alloy}(0)|^2$ for Au in AuNi and CuAuNi. We will discuss ΔE for CuAuNi below.

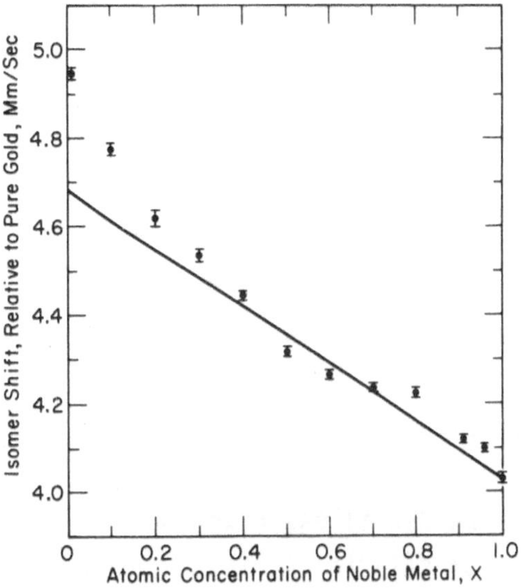

Fig. 1. Isomer shift for $Cu_{x-0.01}Ni_{1-x}Au_{0.01}$ Alloys.

Within available information, it is reasonable to expect the order parameters α_i of Eq. 7 to be small.[14] We shall assume $\alpha_i = 0$ for all i. The atomic volume, V, as a function of x has been measured for Cu_xNi_{1-x}. Thus we can calculate $(V_{Au}/V)^\gamma$. The c_i are known since the alloys have an average f.c.c. structure.

Approximately, for the Ni alloys, there will be a Friedel oscillation for each spin projection, ↑ and ↓. The Au isomer shift, Eq. 7, will be related to the sum of these charge oscillations, i.e., to $\rho_{Ni}(r_i) = \rho_{Ni\uparrow}(r_i) + \rho_{Ni\downarrow}(r_i)$. We have no direct information about $\rho_{Ni}(r_i)$ but we will obtain an estimate of it from Eq. 8.

We note that Ni and Cu are adjacent in the periodic table and that their model potentials will be roughly similar. To obtain an estimate of $\rho_{Ni}(r_i)$ we will take their model potentials to be the same, i.e., $\rho_{Ni}(r_i) = \rho_{Cu}(r_i)$ where the latter is given by Eq. 8. With these assumptions we may calculate ΔE for Au in the AuNi and CuNiAu alloys. The solid curve of Fig. 1 is the result of this calculation for CuNiAu. The agreement between this model and the experimental data is within 2% over most of the range of x

ρ_{Ni} will not exactly equal ρ_{Cu}. The agreement shown does, however, serve to suggest a systematic, similar behavior of the model potential Eq. 4 for these two elements. The model gives $\Delta E(x)$ rather closely in magnitude and in particular it gives the correct sign for ΔE for Au in these Ni alloys even though the AuNi and CuAuNi alloys are strongly magnetic but the CuAu alloys are not.

This model describes how $|\psi_{alloy}(0)|^2$ for Au behaves in the Ni alloys, but does not, of itself, show why the alloy magnetization decreases with increasing noble metal content for $x < 0.6$. The magnetic behavior may be associated primarily with correlation effects in the d-shell, and may not have any simple or explicit dependence on charge transfer apart from the requirements of the Friedel sum rule, i.e., charge neutrality.

The Recoilless Fraction

(e) Measurements of f(T) for Au in the range $4.2 \leq T \leq 100°K$.

In Fig. 2 we show two Mossbauer spectra, one for a thin sample and one for a thick sample of pure Au.

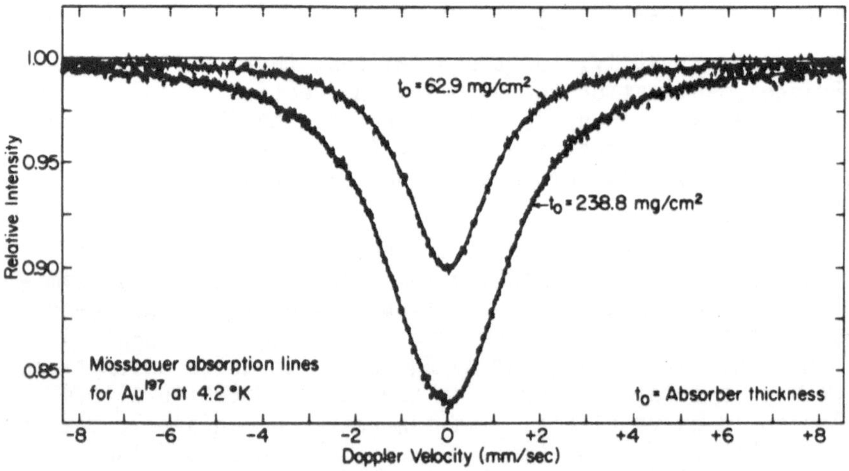

Fig. 2. Mossbauer spectra for two gold samples.

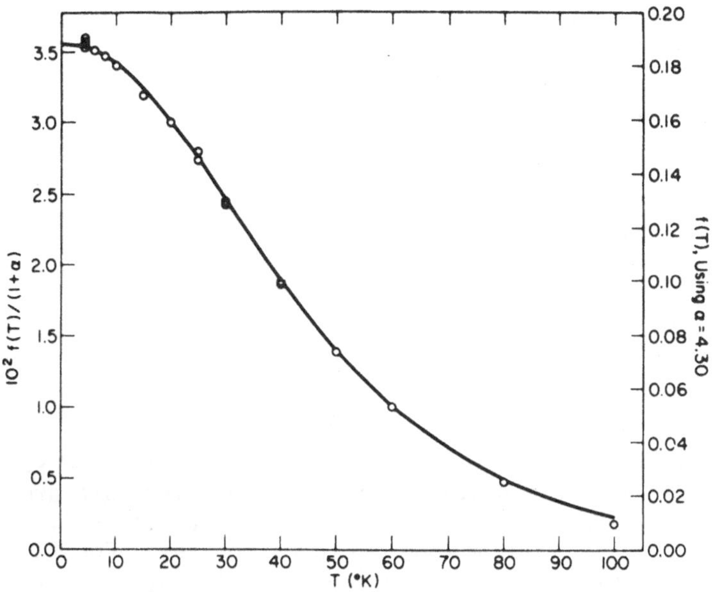

Fig. 3. Recoilless fraction as a function of temperature, f(T), for Au. α is the internal conversion coefficient.

The width of the line is greater for the thicker sample and the line shape thus contains cross section information.

We have developed[3] a least squares program which fits the exact theoretical thick absorber line shape to a Mossbauer spectrum. The solid curves through the data points of Fig. 2 are the results of fitting this exact theoretical line shape function to the data.

From this fit of the true theoretical line shape to line shape measurements for two samples of quite different thickness, it is possible, by an iterative procedure, to measure the natural width Γ_0 and $f/(1\pm\alpha)$ where f is the recoilless fraction. This has been done[3] for Au. Measurements of the line shape were made for thick samples as a function of temperature for $4.2 \leq T \leq 100°K$. Through a comparison of the measured $f(T)/(1\mp\alpha)$ with a theoretical treatment of the Debye-Waller factor by Skelton and Feldman,[18] and by an analysis of nuclear data, a value for α was obtained. The results[3] are $\Gamma_0 = 0.923 \pm 0.006$ mm/sec., $\alpha = 4.30$ and Fig. 3 shows f(T) and $f(T)/(1+\alpha)$ for Au.

(f) Measurement of f(T) for Au as a function of temperature and of short range order for Cu_3Au.

Given α, we can reduce a measurement of the Au Mossbauer cross section to obtain f(T) for Au in an alloy. This has been done for Au in Cu_3Au as a function of temperature, $4.2 \leq T \leq 100°K$, and of short range order for an ordered and for a disordered phase.[19] f(T) is about 6% larger for the ordered than it is for the disordered phase. A theoretical calculation of f(T) for completely ordered and for totally random phases of Cu_3Au has been made by Kvashnina and Krivoglaz.[20] The theoretical and experimental results are in approximate agreement.

(g) Measurements for f(T) for an alloy of two percent Au in Cu.

In Fig. 4 we show the results[21] of our f(T) measurements for a dilute solution of Au in Cu. The data are compared with the theoretical models of Dawber and Elliott[22] (lower curve) and of Mannheim[23] (upper curve). For the Dawber and Elliott model applied to Au in Cu, a Au atom replaces a Cu atom in the Cu lattice. There is a change of mass from M_{Cu} to M_{Au} at the gold site, but the force

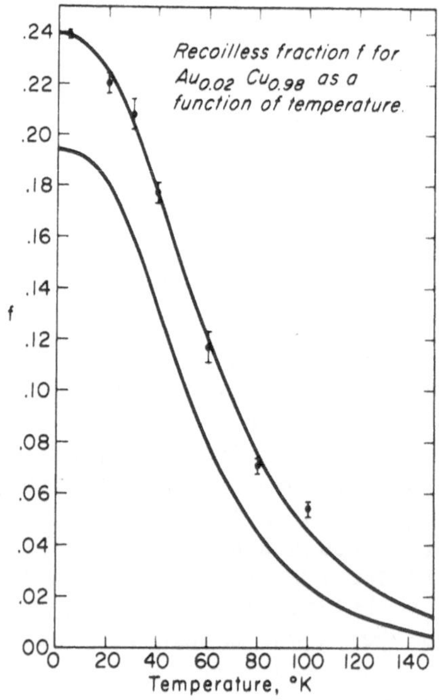

Fig. 4. Recoilless fraction f(T) for a 2 percent solid solution of Au in Cu.

constants holding the Au are left the same as those binding the original Cu atom in pure Cu. In the Manhein model, the mass M_{Au} is again substituted for M_{Cu} and the nearest neighbor force constants are also allowed to change by a single adjustable factor. We have adjusted this parameter to obtain a fit of the Manheim model for f(T) to our data, Fig. 4. This model gives the result that the nearest neighbor force constants binding Au in Cu are 1.48 $\pm$ 0.03 times greater than these force constants for pure Cu.

Finally, we note that it should be possible to obtain a description of f(T) for Au in an alloy in terms of the potential W of Eq. 4 and thus correlate f(T) with ΔE, $\Delta\rho/c$ and other alloy properties.

References

1. Louis D. Roberts and J. O. Thomson, Phys. Rev. 129, 664 (1963).

2. Louis D. Roberts, R. L. Becker, F. E. Obenshain and J. O. Thomson, Phys. Rev. 137, A 895 (1965).

3. D. J. Erickson, Louis D. Roberts, J. W. Burton and J. O. Thomson, Phys. Rev. B, 3, 2180 (1971).

4. H. Frauenfelder, The Mossbauer Effect (W. A. Benjamin, Inc., New York, 1962).

5. C. Kittel, Quantum Theory of Solids (John Wiley and Sons, New York, 1967).

6. J. Friedel, Nuovo Cimento, 7, 287 (1958).

7. A. Blandin and E. Daniel, Phys. Chem. Solids, 10, 126 (1959).

8. E. Daniel, Phys. Chem. Solids, 10, 174 (1959).

9. V. Heine and I. Abarenkov, Phil. Mag. 9, 451 (1964).

10. W. D. Josephson, D. W. Knoble and Louis D. Roberts, to be published.

11. Louis D. Roberts, D. O. Patterson, J. O. Thomson, and R. P. Levey, Phys. Rev. 179, 656 (1969).

12. Thomas C. Tucker, Louis D. Roberts, C. W. Nestor, and Thomas C. Carlson, Phys. Rev. 178, 998 (1969).

13. Paul G. Huray, Louis D. Roberts, and J. O. Thomson, Phys. Rev. B 2, 2440 (1970).

14. J. W. Burton, J. O. Thomson, P. G. Huray and Louis D. Roberts, Phys. Rev. in press.

15. E. C. Stoner, Phil. Mag. [7], 15, 1018 (1933).

16. E. P. Wohlfarth, Proc. Roy. Soc. A195, 434 (1948).

17. J. W. Cable, E. O. Wollan, and H. R. Child, Phys. Rev. Letters, 22, 1256 (1969).

18. E. F. Skelton and J. L. Feldman, Acta. Cryst. A27, 484 (1971).

19. D. J. Erickson and Louis D. Roberts, Phys. Rev. in press.

20. L. B. Kvashnina and M. A. Krivoglaz, Phys. Metals Metall. 25, 1 (1968).

21. J. F. Prince, Louis D. Roberts and D. J. Erickson, Bull. Am. Phys. Soc. 17, 292 (1972).

22. P. G. Dawber and R. J. Elliott, Proc. Roy. Soc. (London) 273A, 222 (1968).

23. P. D. Mannheim, Phys. Rev. 165, 1011 (1968).

*Research jointly sponsored by the U.S. Atomic Energy Commission under Contract No. AT-(40-1)-3897 with the University of North Carolina, and by the Advanced Research Projects Agency under Contract DAHC-15-67-C-0223 with the University of North Carolina.

**Present Address: Los Alamos Scientific Laboratory, Los Alamos, New Mexico 87544.

MÖSSBAUER EFFECT STUDIES OF PHASE TRANSITIONS

G.K. Shenoy

Physik-Department

Technische Universität München

D-8046 Garching, Germany

The use of the Mössbauer effect as a tool to investigate phase transitions is discussed. While the emphasis thus far has been on the study of magnetic phase transitions, in recent years many other transitions, such as displacive, band Jahn-Teller, cooperative Jahn-Teller, ferroelectric, metal-insulator, order-disorder transitions, have also been investigated using the Mössbauer effect. Through examples some of this work is here reviewed.

1. INTRODUCTION

The existence of phase transitions has been known for almost 50 years; however, only during the past decade has there been a renewed interest in transition phenomena, particularly in the behaviour of the physical systems at critical points /1-3/. A wide variety of systems undergo phase transition at a critical temperature, a critical pressure, or a critical composition; and magnets, alloys, and superconductors are just a few examples. The functional dependence of physical quantities like the magnetization, specific heat or the dielectric constant in the critical region are usually used to observe the phase transitions as well as to verify various critical relations. While in earlier times the behaviour of bulk properties near the phase transition was of major interest, the emphasis has shifted in recent years to an investigation of the microscopic properties. Techniques such as NMR, EPR, PAC and the Mössbauer effect (ME) have thus come into the foreground. This paper is concerned with the observation of phase changes and the study of critical regions using the ME.

The present emphasis will be on experiments and their understanding. While a wide variety of phase transitions have been investigated using the ME, in this paper we shall deal only with those transitions (Table I) that we consider most important. Therefore this review makes no attempt to provide a complete reference work on the subject.

2. MÖSSBAUER EFFECT AS A TOOL TO STUDY PHASE TRANSITIONS

The magnetic phase transition (PT) is perhaps the easiest one to observe by the ME through a study of the hyperfine (hf) magnetic field at the nucleus as a function of temperature. In the simplest case, the hf magnetic field is zero above the transition temperature, and below the transition temperature the hf magnetic field appears gradually and can be observed by the nuclear Zeeman splitting. Complications may arise through the observation of 'relaxation spectra'. Structural phase transitions are often associated with an unstable 'soft' optical mode. Hence they might be observed either through changes in the quadrupole interaction (ΔE_Q), the centre shift (CS) of the Mössbauer spectrum, or the Mössbauer resonance fraction (f), when measured as a function of temperature. However, all these quantities are less sensitive to the transition than the magnetic properties and extreme care has to be exercised in their evaluation. The metal-insulator transitions often produce a measurable change in the isomer shift (IS).

Observing the PT is one aspect of these studies. Many new PT have been established by the ME. A few examples are the magnetic transitions in ortho-

Table I

Transition	Example	Reference
Ferromagnetic	Fe metal	4
	FeNi	5,6,7
	$NpAl_2$	8
	EuO	9,10
	Eu metal	11
Antiferromagnetic	NpS	12
	Iron fluorides	13
	$REFeO_3$	14
	$HoFeO_3$	15
Structural		
a. Displacive	$SrTiO_3$	16
b. Band Jahn Teller	Nb_3Sn	17
c. Cooperative	$CuFe_2O_4$	18
Jahn Teller	Spinels	19
	$FeCr_2O_4$, FeV_2O_4	20
d. Ferroelectric	$BaTiO_3(Fe)$, $PbTiO_3(Fe)$ $BaTiO_3(Sn)$	21-23
	$K_4Fe(CN)_6 \cdot 3H_2O$	24-28
	$NH_4Fe(SO_4)_2 \cdot 12H_2O$	25
e. Martensitic	REAu	29
f. Others	$Fe(bcc) \rightarrow Fe(fcc)$	4
	$Fe(bcc) \rightarrow Fe(hcp)$	20
	$Fe(ClO_4)_2 \cdot 6H_2O$	31
Hindrance of Molecular Motion	Fe(II) hexammines	32
	Fe Polyacrylonitrile	33
Metal-Insulator	$V_2O_3(Fe)$, $(V_{1-x}Cr_x)_2O_3(Fe)$	34
Liquid-Glass	Sn and Fe systems	35
Crystalline-Amorphous	Sb	36
	Se(Te)	37
Order-Disorder	$FeNi_3$	38

pyroxenes /30/ and in Eu and Gd compounds /40,41/, the structural transitions in $Fe(ClO_4)_2 \cdot 6H_2O$ /31/ and $Fe(py)_2Cl_2$ /42/ and the transitions accompanying hindrance of internal motion in Fe(II) hexammines /32/ and $FeSiF_6 \cdot 6H_2O$ /43/. Even more interesting (and also more difficult) is the investigation of the sensitive quantities throughout the whole critical region. (Experimentalists usually consider the region $0.9T_c \leq T \leq T_c$ to be the critical region.) The study of the critical region is normally associated with the deduction of a critical exponent. The physical quantity (say specific heat, or the hf magnetic field) can be expressed as a function of temperature T by the expansion:

$$\emptyset(\varepsilon) = A\varepsilon^{x_1} (1+B\varepsilon^{x_2} + \dots) \qquad (2\text{-}1)$$

where $\varepsilon=(T-T_c)/T_c$, T_c being the critical temperature. Then the critical exponent is defined by

$$\lambda = \lim_{\varepsilon \to 0} \frac{\ln \emptyset(\varepsilon)}{\ln \varepsilon} \qquad (2\text{-}2)$$

Obviously the critical exponent contains less information than the functional form $\emptyset(\varepsilon)$. However, sufficiently close to the critical point the leading term dominates so that the determination of the critical exponent from a ln-ln plot may retain most of the information through $\lambda = x_1$. More important is the fact that there exist a number of relations between the critical exponents of various physical quantities which in turn define the equation of state of the solids investigated. Hence determinations of the exponents by different techniques form a major effort in the understanding of critical phenomenon.

Finally, a few remarks should be made regarding general experimental considerations for the ME technique. Needless to say good samples are of crucial importance in phase transition studies. They should be of the highest purity and be free of defects - particularly for critical region studies. It must be realized that variations in material properties throughout the sample will result in a smearing out of the critical point; yet this spread has to be confined to a region extending to not more than a few percent of the whole critical region being investigated. Temperature inhomogeneities in ME absorber studies are common because of the large absorber areas needed for this technique. They can be reduced by using isothermal enclosures or by reducing the absorber area to 1 to 2 mm^2 /44/. More dependable results can be obtained through temperature scanning techniques /5/, and the use of small computers as multiparameter storage devices. Finally, it should be mentioned that the mechanical clamping of the sample may lead to erroneous results, particularly in the study of PT involving large lattice changes (e.g. in ferroelectrics and frozen solutions). These aspects and other experimental considerations have been discussed in detail elsewhere /3,4,7,13,15/.

3. MAGNETIC TRANSITIONS

a. Hyperfine Magnetic Field

The critical region in ferro- and antiferromagnets has been extensively studied, both theoretically /1,2/ and experimentally /3/. A variety of bulk properties such as the magnetization (or sublattice magnetization), specific heat, and susceptibility have been measured to probe the critical region. They are found to follow a power law with a specific critical exponent (Table II). In micorscopic techniques one measures the value of the hf magnetic field at the nucleus as a function of temperature. In first order this field, $H_n(T)$, is considered to be proportional to the local magnetization. This leads one to assume that the value of $H_n(T)$ is proportional to M(T) the spontaneous magnetization. Quite often M(T) arises from several sources of magnetism (e.g. spin, orbital, and conduction electron polarization). Then the above proportionality depends on whether the ratios of the different contributions show a temperature dependence. Experimentally it has been found, for example that in Fe metal M(T) and $H_n(T)$ are nearly proportional over the entire temperature range /4/.

Table II

Quantity	Exponent	Value for 3 dimensional Heisenberg magnet	Power law	conditions
Specific heat at	α	0	$\varepsilon^{-\alpha}$	$\varepsilon>0$
fixed field	α'	0	$(-\varepsilon)^{-\alpha'}$	$\varepsilon<0$
zero field isothermal	γ	4/3	$\varepsilon^{-\gamma}$	$\varepsilon>0$
susceptibility	γ'	4/3	$(-\varepsilon)^{-\gamma'}$	$\varepsilon<0$
zero field magnetization	β	1/3	$(-\varepsilon)^{\beta}$	$\varepsilon<0$

Hence we can write

$$\frac{H_n(T)}{H_n(0)} = B(-\varepsilon)^{\beta} \qquad (3-1)$$

The factor B is dependent for example on the spin wave spectrum of the solid. The critical exponent, β, is expected to be nearly the same for all magnets with the same dimensionality in ordering; the hf magnetic fields have been measured in many systems close to their critical points to test this explanation.

In deducing β from the function $H_n(T)$, two techniques have been followed. One is to plot $\ln(H_n(T)/H_n(0))$ as a function of $\ln(\varepsilon)$. This procedure presupposes the exact knowledge of the critical point and hence may not be reliable. The other is to plot the deviation, $\Delta T/T_c$, from the linearity of a $(H_n(T)/H_n(0))^{1/\beta}$ plot for various values of β.

Thus

$$\frac{\Delta T}{T_c} = -\varepsilon - (H_n(T)/BH_n(0))^{1/\beta} \qquad (3\text{-}2)$$

has a minimum deviation in the critical region for the best choice of β. The details of this approach are given in /13,45/. Neither of these techniques is fully satisfactory. At the present perhaps the best way to deduce β, B, and T_c is to least-squares fit the experimental data to Eq. (3-1) by allowing all three quantities to vary freely. It is however essential to weigh the values of both $H_n(T)$ and T through the introduction of the appropriate individual error values within the fitting procedure. Even this method will give reliable results only if data of very good quality are available.

Table III

System	T_c(K)	Order	β	Ref.
Fe metal	1042.0	II	0.342±.004	47
			0.37 ±.02	47
FeNi	629.4	II	0.33 ±0.03	5
			0.38 ±0.01	6
			0.378±0.040	7
EuO	69.3	II	0.34 ±0.02	9
			0.36 ±0.03	10
$NpAl_2$	55.8	II	0.335±.017	8
FeF_3	363.12	II	0.352±.006	13
$RbFeF_3$	102.0	II	0.333	13
$REFeO_3$		II	~0.345	14
Eu	89.5	I	-	11
NpS	19.0	I	-	12
Anhydr.$FeCl_3$	8.7	II	0.146	47
$KFeF_4$	141.51	II	0.209±.008	48
3 dimensional Ising	(theory)	II	0.313±0.004	49
2 dimensional Ising	(theory)	II	0.125	50
Molecular field	(theory)	II	0.5	51
Landau Green Function	(theory)	II	0.5	51
RPA	(theory)	II	0.3334	52
Callen Decoupling	(theory)	II	0.334	52
Two Spin Cluster	(theory)	II	0.334	52

In Table III, the values of the critical exponent β for some of the magnetic systems investigated by the ME are listed. Most of these results are close to the value $\beta = 0.313$ predicted for a 3 dimensional Ising ferromagnet /49/. However, the hf magnetic fields in anhydrous $FeCl_3$ /47/ which have been measured recently near its Néel temperature give a value of $\beta = 0.146$. This value is close to that predicted for a magnet with two dimensional ordering, viz. $\beta = 0.125$. Thus it is conjectured that there may be a nearly two dimensional magnetic ordering in this layer-structured material. Recently, $KFeF_4$ has also been established to be a planar antiferromagnet through both neutron scattering measurements and the deduction of β from Mössbauer hf magnetic-field measurements in the critical region /48/.

If the Mössbauer atom is incorporated as an impurity into a magnetically ordered host lattice, the dependence of $H_n(T)$ need not be proportional to M(T). In the Fe$\underline{\text{Ni}}$ system, $H_n(T)$ at the impurity has been found to closely follow the host magnetization. On the other hand, the field at Fe nucleus in $MnYO_3$ largely deviates from M(T) /53/. An extreme example of such a non-proportionality is the hf magnetic field at Mn nucleus in Fe host (investigated by NMR) /54/. The question of proportionality clearly depends on the strength of the exchange coupling between the impurity spin and host spins /55/. A modified Brillouin function might then be used to connect the hf magnetic field at the impurity with the host magnetization:

$$(H_n(T)/H_n(0)) = B_J\ (\zeta\ T_c M(T)/TM(0)) \qquad (3\text{-}3)$$

where $\zeta = g\ J\ H_{ex}^{imp}(0)/kT_c$, $H_{ex}^{imp}(0)$ being the exchange field acting on the impurity. For the Mn$\underline{\text{Fe}}$ system $\zeta = 0.75$ and $J = 3/2$ /53/ and for the Fe$\underline{\text{Ni}}$ system $\zeta = 2.3$ and $J = 3/2$ /3/. If Eq. (3-3) describes $H_n(T)$ accurately in the vicinity of the transition point, then it is possible to obtain M(T)/M(0) as a function of the temperature in the critical region; and thus to deduce the β of the host in turn. However, the reliability of such a complex procedure may be questionable.

It should be pointed out that the experimental determination of hf magnetic fields close to a critical point is not always possible. In the cases discussed above the relaxation time of the electron spin τ_R is short compared to the Larmor precession time of the nuclear spin, τ_L. There are many examples where this is not the case. If $\tau_R \gg \tau_L$, then one observes a completely split hf pattern even in the paramagnetic range, and the transition into the magnetically ordered region will not be noticeable in the Mössbauer pattern /56/ - unless the exchange field changes the character of the electronic ground state /57/. In either case a detailed investigation of the critical region may be uninformative.

There exist numerous theoretical investigations /58/ of the fluctuation of the electronic spin and of its critical slowing down in the immediate neighbourhood of the critical point. The dynamical scaling laws describe the spin relaxation time by

$$\tau_R = c \cdot (\varepsilon)^{\delta}$$

From the measurement of the line width slightly below and above T_c for the FeNi system δ has been found to be -0.5 ± 0.1 for $T < T_c$ and -1.0 ± 0.1 for $T > T_c$ /6/, where the latter is in agreement with theory /58/. Groll /9/ finds similar behaviour in EuO.

There are also reports of "anomalous" spectra close to the ordering temperature /59/. These spectra consist of a magnetically split pattern and a single-line paramagnetic pattern simultaneously coexisting slightly below the transition point. These patterns have been interpreted as arising from a collective spin fluctuation near the critical point resulting in superparamagnetic domains /59/.

In the discussion thus far it had been tacitly assumed that the PT investigated are of second order. Magnetic PT of first-order are also observed by the ME. They are characterised by a sudden drop of $H_n(T)$ from finite value to zero at T_c. The observation of a coexistence of the paramagnetic and the magnetically ordered phase over a small temperature region close to T_c is also typical. Examples are Eu metal /11/ and NpS /12/.

b. Centre Shift and Mössbauer Fraction

The position of the centre of the Mössbauer spectrum can show an anomalous change across the PT. This could be due either to the changes in the second-order Doppler shift (SOD) or the isomer shift (IS), or both. Bashkirov and Selyutin /60/ have obtained an expression for the SOD in a crystal containing exchange-coupled atoms. Utilizing a Debye picture for the vibrational frequency spectrum of the solid, they obtain

$$SOD = - \frac{9kTE_\gamma}{4Mc^2}\left[\frac{1}{4}\frac{\Theta'_D}{T} + 2\left(\frac{T}{\Theta'_D}\right)^3 \int_0^{\Theta'_D/T} \frac{x^3\,dx}{e^x - 1}\right] \quad (3\text{-}4)$$

with $\Theta'_D = \Theta_D\,(1+B)^{1/2}$ being the Debye temperature for a magnetically ordered solid and

$$B = - 8\pi^2 \quad (\vec{S}_i \cdot \vec{S}_j)a^2/v^2M \quad J''(a) \quad (3\text{-}5)$$

where a is the lattice parameter, v is the velocity of sound, and J"(a) is the second derivative of the exchange integral. In a molecular field approximation B is proportional to $(M(T)/M(0))^2$.

The variation of the IS is proportional to the changes in the total electron density at the nucleus accompanying the PT. In any real case this dependence is far too complicated to be evaluated theoretically. A few general remarks will have to suffice. In a magnetic insulator both covalency and overlap effects should change at the PT, thus changing the electron density. In a conductor it is essential to consider the influence of magnetic order on the band structure /61,62/. In addition, any anomalous volume changes (e.g. magnetostriction) in the lattice at the PT can affect both the SOD and the IS /62,63/.

The observed change of about 0.03 mm/sec in the CS at the ferromagnetic transition of Fe metal /64/ has been interpreted in terms of a variation of the IS /61,62/. The effect is considered to arise from a shift in the absolute position of the Fermi level located within the s-band, caused by the splitting of the d-band by the magnetic interaction. Ingalls /62/ elaborated on this picture by performing a calculation of the shape of the s-band. Muller /65/ investigated the band structure for Fe metal with a high accuracy integration procedure and found a sharp structure in the s-like density of states. Extending these calculations /66/ to examine the effect of magnetic ordering, it was observed that with the onset of a molecular field the Fermi level moves through this structure in the s-like density of states which accounts for the observed change in the IS.

A change in the CS has also been observed in certain insulators. The changes observed in FeF_3 /67/ and in $HoFeF_3$ /15/ have been accounted for largely through SOD changes using Eq. (3-4). A jump in the CS of about 0.1 mm/sec has been observed at the transition in $KFeF_4$ /48/.

It is evident from the above discussion on θ'_D and from Eq. (3-5) that the Mössbauer resonance fraction f should also show some dependence on the magnetization of the lattice. There are numerous experimental reports concerning this, although an exhaustive study has not yet been undertaken. Recently, Nandwani and Puri /68/ have calculated in detail the temperature dependence of f in metallic Fe /69,70/. They utilized the experimentally determined phonon frequency spectrum, and included an anharmonic term as well as a term proportional to the magnetization. They noted that the inclusion of the last term in particular improves the fit to the low temperature data /69,70/.

c. Quadrupole Interaction

The onset of magnetic ordering can induce a quadrupole interaction even in those cases where the symmetry of the atom is cubic. If the orbital electrons contribute to the magnetism, the magnetic ordering aligns the orbital momentum. This results in a non-spherical charge distribution producing an electric field gradient (EFG). Such an interaction has been observed in many 3d, 4f, and 5f systems /71,72/. In addition, departures from strictly cubic symmetry due to magnetostriction can produce an additional EFG in the magnetically ordered state as in the case of Fe metal /73/.

4. STRUCTURAL PHASE TRANSITIONS

There have been a few investigations of structural phase transitions using the ME. As stated earlier, such a PT can be detected in a Mössbauer spectrum, through a variation of ΔE_Q, f, or CS. In general one or more of these parameters change at T_c. It is impossible to give here either a complete account of the theoretical nature of structural PT (for details see /74/) or even a general treatment of the change in the Mössbauer parameters accompanying a structural transition. It has however become customary to broadly divide structural PT into three groups: a) displacive, b) cooperative Jahn-Teller, and c) band Jahn-Teller transitions.

a. Displacive Transitions

Most of the ferroelectric transitions are typical examples of displacive transitions. In addition such transitions are exhibited by certain perovskite structured compounds (like $SrTiO_3$ and $LaAlO_3$) which are not ferroelectrics. The general theory of displacive transitions has been given by Cochran /75/. The essentials of this theory can be described by considering the motion of an ion in a lattice using a single oscillator model. The forces acting on this consist of i) a short range harmonic restoring force, F_s, ii) a long range electrostatic driving force, $-F_l$, which tends to drag the ion away from the equilibrium, and iii) an effective anharmonic restoring force, BT, which is temperature dependent. The oscillator frequency is then

$$\omega_o^2 = \frac{B}{T}(T - T_o) \tag{4-1}$$

where

$$T_o = \frac{F_l - F_s}{B} \tag{4-2}$$

Usually, T_o is referred to as the Curie-Weiss temperature. In order to have a displacive transition, it is necessary that some lattice modes vanish or that $\omega_o^2 < 0$. In ferroelectrics $|F_l|>|F_s|$ and below T_o one obtains a displacement of the ion and the vanishing of a mode. However, even without the presence of long range forces, F_l, a phase transition could take place if the short range forces were negative. This is the situation in $SrTiO_3$ for example.

The $SrTiO_3$ lattice has been probed by introducing ^{57}Co which decays to Fe^{3+} at the Ti^{4+} position. The phase transition occurs at 110 K, where the lattice changes from a cubic to a low-temperature tetragonal phase /74,76-78/. In the Mössbauer studies /16/ no measurable quadrupole interaction was found in the tetragonal phase at the Fe^{3+} nucleus. This result is not unexpected since c/a = 1.0005 /78/ below 110 K. There is however a change in the CS at the transition.

Ferroelectric PT have drawn considerable attention during recent years. Reports from different groups regarding the observation of an anomalous temperature dependence of f in some of the ferroelectrics have been conflicting. As discussed above, a cancellation of long-range Coulomb forces by short-range forces leads to these displacive transitions and also to an anomaly in the frequency of a transverse optical mode /75/ in the crystal. Muzikar et al. /79/ suggested that this anomaly of ω_o^2 should lead to a minimum in f. Gleason and Walker /25/ and Bhide and He /22/ have given an explicit expression for this parameter in a ferroelectric:

$$f(T) = f_D(T)\exp\left\{ - \left(\frac{2}{e^y - 1} + 1\right) \left[\frac{(\hbar K)^2}{2M\hbar} a_T^2 \right] \frac{\hbar}{kT} \frac{1}{y} \right\} \quad (4\text{-}3)$$

where

$$y = \frac{\hbar}{kT} \quad G(T - T_o)^{1/2} \quad (4\text{-}4)$$

Here $f_D(T)$ is the normal f factor derived from a Debye spectrum of the crystal without the anomalous phonon mode; K is the wave vector of the gamma ray; M is the mass of the atom; a_T^2 is the coefficient of the normal coordinates for the anomalous mode; and G = B/M as defined in Eq. (4-2).

For a second-order ferroelectric transition $T_o = T_f$ where T_o is the Curie-Weiss temperature given by Eq. (4-2) and T_f is the ferroelectric phase transition temperature. Thus at $T = T_f$, $y = 0$ in Eq. (4-3) which results in a vanishing of the Mössbauer effect at the transition. For a first-order transition $T_o \neq T_f$, and hence y never equals zero. However, in that case y has a minimum at $T = T_f$ which results in a minimum in f(T) as well. All these arguments are of course true only in a simple lattice; the strength of the anomaly in f(T) in a real crystal is dependent on many details such as the phonon spectrum, the damping of anomalous mode and anharmonic effects.

Such anomalies in f(T) have been reported in $BaTiO_3$ /21/ and $PbTiO_3$ /22/. ME experiments were performed by introducing into these lattices ^{57}Co which decays to Fe^{3+} at the Ti^{4+} site. The main problem in such sources is to achieve a charge compensation in the lattice far away from the Mössbauer probe. These difficulties are absent in Sn^{4+} doping experiments performed on $BaTiO_3$ /80/. There are conflicting reports regarding $K_4Fe(CN)_6 \cdot 3H_2O$ /24-28/ and it now appears that the quality of the single crystal is of primary consideration in this case. Pronounced f(T) anomalies have also been observed in ferroelectric $NH_4Fe(SO_4)_2 \cdot 12H_2O$ /25/ and in iron- and nickelboracites /81/.

In all the perovskite ferroelectrics ($BaTiO_3$, $PbTiO_3$) a large quadrupole interaction has been observed in the ferroelectric phase with a tetragonal structure and this reduced to zero above the ferroelectric transition temperature where the lattice is cubic /21,22/. In $BaTiO_3$ it is found that ΔE_Q is proportional to the square of the spontaneous polarization, the tetragonal strain in the lattice, as well as to the birefringence /21/. Similar observations have been made in $KNbO_3$ /82/. These results are consistent with the displacive transition theory.

In most of the cases mentioned above an anomalous change in the CS has also been reported at the ferroelectric transition. Since the change is a result of variations in both the SOD and the IS, the detailed understanding of the changes has remained only in a speculative state /22/. The IS changes are due to both changes in the lattice volume and in the nature of the chemical bonds. The SOD will be affected by the difference in the Debye temperature above and below the PT and also by the vanishing of the transverse optical mode. Detailed correlations with other experimental techniques are needed for any improvement in the understanding of these anomalies.

Finally it should be mentioned that the f(T) anomaly in antiferroelectrics should be even larger than in ferroelectrics /83/. This is due to the fact that the energy of the phonons belonging to the end of the optical transverse branch is anomalously low and strongly temperature dependent. No experiments to date have been reported on antiferroelectrics.

b. Cooperative Jahn-Teller Transitions

The symmetry of a non-linear molecule is often not stable if its electronic ground state is orbitally degenerate. The stable form is then one of lower symmetry in which all orbital degeneracy is lifted. For example, a complex molecule with a d^1 configuration of the central ion should exist neither in an octahedral nor in a tetrahedral environment. This is known as the Jahn-Teller effect /84/. If a solid is made up of ions which are all in such an unstable configuration, a crystallographic phase transition may occur at some temperature. The distortions produced at individual sites interfere with one another and lead to an alignment of distortions to produce macroscopic changes in the crystal symmetry. This phenomenon is called the cooperative Jahn-Teller effect. Such transitions have been observed using the ME in iron-based oxides such as

$CuFe_2O_4$ /18/, $FeCr_2O_4$, FeV_2O_4 /20,85/, and $Ge_{0.2}Cu_{1.2}Fe_{1.6}O_4$ /19/.

The work on $FeCr_2O_4$ and FeV_2O_4 indicates a quadrupole interaction at the tetrahedral Fe^{2+} site even above the transition temperatures of 135 K and 127 K, respectively, although at room temperature a single resonance was observed. Near the transition temperature the lines broadened resulting finally in a sharp increase in the quadrupole splitting. Hartmann-Boutron /86/ has accounted for the shapes of these spectra in terms of a dynamic Jahn-Teller effect. Her model involves certain assumptions on jumps between two potential wells for the ion and a temperature dependent jump frequency. There is perhaps no need for such an elaborate model. The transition may simply be sluggish and therefore the distortion continues to increase even below the transition temperature /20/.

c. Band Jahn-Teller Transitions

Numerous systems undergo structural transitions between higher (usually cubic) to lower structural symmetries. A typical example is Nb_3Sn which has been investigated by Shier and Taylor /17/ using the ME in ^{119}Sn. The phase transition which occurs around 43 K is considered to be a band analogue of the localized Jahn-Teller effect /87/ discussed above. This material has a nearly full triply degenerate d band at k = 0 in the cubic state. The Jahn-Teller distortion leads to a lifting of the degeneracy. The energy involved in the process of occupation of the split d-band is balanced by changes in phonon-mode energies. As a consequence one might expect to observe a quadrupole-split spectrum below the transition temperature, a change in the CS accompanying the redistribution of band electrons, and an anomaly in the f(T) due to vanishing of certain 'soft' phonon modes. Indeed, the authors of /17/ report anomalous behaviour of both the CS and the f(T) with temperature. However, the results are not conclusive since the anomalies are small and are observed even above 43 K. No quadrupole interaction was observed at low temperatures perhaps due to the small distortion (c/a = 1.0062) involved in this phase transition.

5. HINDRANCE OF MOLECULAR MOTION

The phase transition accompanying a hindrance of molecular motion is a phenomenon which has been well studied by NMR, EPR and specific heat techniques. In iron hexammines this phenomenon was first observed through Mössbauer studies, by Asch, et al. /32/.

In the hexammines of transition elements there is a group of six NH_3 units which rotates around the transition element to produce a cubic crystal field at the ion. At lower temperatures the motion slows down and finally below a critical temperature the molecule goes to a lower symmetry through a cooperative electrostatic interaction. A detailed theory has been given by Bates and Stevens /88/ who consider the electrostatic interaction between various NH_3 groups and predict a sharp critical point. The theory expalins the EPR investigations /89/ as well as the specific heat measurement /90/ on Ni hexammines.

The Mössbauer spectra of most of the Fe(II) hexammines show a single line at room temperature which abruptly splits into a quadrupole doublet below a well-defined temperature. This transition temperature is a function of the other ligands in the salt. The effective cubic symmetry of Fe(II) ion above the transition results from a free motion of the NH_3 ligands. When the motion is hindered, the cooperative phenomena reduces the symmetry of the lattice to a trigonal one. Consequently, a quadrupole pattern is observed.

In $FeSiF_6 \cdot 6H_2O$, the iron is surrounded by a $6H_2O$ octahedron. Around 195 K, the motion of the protons between two possible positions slows down and a trigonal distortion is produced. This has been observed through an anomalous change in the quadrupole interaction /43/. An additional aspect of these phase transitions is that in both the Fe hexammines /91/ and $FeSiF_6 \cdot 6H_2O$ /43/ there is an abrupt change in f(T) at the transition temperature.

The freezing-in of rotational degrees of freedom in a polymer has recently been reported as the polymer goes to a glassy state /33/. ^{57}Fe has been incorporated in a polyacrylonitrile (PAN) and the f(T) carefully measured as the polymer becomes glassy. The transition is marked by an abrupt change in the temperature dependence of f(T). Various other liquid-glass transitions have been discussed by Ruby /35/ at this conference.

6. METAL-INSULATOR TRANSITIONS

Recently, considerable interest has been shown in metal-insulator PT induced either by the variation of the pressure or the temperature. Ge, Si, Se, InAs, and ZnS are a few examples for the pressure - induced PT, while the oxides of transition elements, such as VO_2, V_2O_3, and Ti_2O_3 exhibit such a PT at a certain critical temperature. Also by varying the composition one can induce a transition from metal to insulator. These transitions are usually observed through the measurement of the conductivity, which changes by 10 to 100 orders of magnitude.

The actual mechanism causing these transitions might be quite different despite the fact that the variation in the electrical conductivity shows rather similar behaviour. The three best known mechanisms /92/ are: 1. Band overlap-, 2. Electronic-, 3. Mott-transition. In band-overlap transition one expects the band gap to shrink through zero as one varies one of the physical parameters. In an electronic transition one needs a narrow-band situation which is unstable towards a crystal transformation or a magnetic transition. For example, Slater /93/ pointed out that an antiferromagnet with a half-filled band above the Néel temperature, T_N, would become an insulator below T_N. In such a situation the metal-insulator transition temperature coincides with T_N. An example is the 160 K transition in V_2O_3. In a Mott transition /94/ the localized electrons become delocalized into a partially filled band. The actual cause of such a transition could be for example the appropriate change in the atomic distance.

All these metal-insulator transitions will basically be noticed in a Mössbauer experiment through changes in the IS. We shall give a few cases where such investigations have been performed. The metal-insulator transition accompanying the precipitation of the antiferromagnetic state in V_2O_3 has been studied by replacing V^{3+} by Fe^{3+} /34, 95/. A sudden jump of about 0.17 mm/sec in the IS has been observed, which is consistent with Slater's picture /93/. Petrich, et al. /10/ have studied the ^{151}Eu resonance in Eu-rich EuO samples which undergo insulator-metal transition close to 50 K, while they order magnetically at 69.3 K. The observed change in the IS (of 0.15 mm/sec) as one goes from below 50 K to above the magnetic ordering temperature has been interpreted in terms of an exchange-induced configurational change (or auto ionization) of two electrons trapped at an axygen vacancy in this defect solid.

$(V_{1-x}Cr_x)_2O_3$ has recently been established as undergoing a Mott transition at 575 K /96/. Wertheim, et al. /34/ have studied this compound by doping with ^{57}Fe. They have explained the observed change in the IS in terms of a return from the metallic to insulating state above 575 K. In brief, the Fe^{3+} forms an acceptor state which is merged with a partially-filled band below 575 K. Above 575 K the change in the IS indicates an increase in d electron localization as expected in a Mott transiton. Wertheim, et al. /34/ also explain the changes in the room temperature IS values reported in $(V_{1-x}Fe_x)_2O_3$ /97/ with varying x, as due to a Mott transition near x = 0.17.

7. OTHER PHASE TRANSITIONS

The phenomenon of phase transition is so common in nature that there are many more examples for which the detailed mechanisms are not well understood. Some of these have been studied by the ME. Both the α-γ and the γ-δ phase transitions in metallic iron have been observed through sudden changes in the CS and the f(T) /4,62,69/. PT has been observed in $Fe(ClO_4)_2 \cdot 6H_2O$ /31/ and in $Fe(py)_2Cl_2$ /42,98/ through large changes in ΔE_Q at the transition temperature. If the quadrupole pattern is asymmetric in intensity (due to either the anisotropic lattice vibrations or the preferred orientation) in one phase, the asymmetry may change in the other phase. There are also many pressure-induced PT which have been observed, as for example the bcc to hcp conversion of iron metal at 130 kbar /30/ and that of Ti metal /99/.

Recently liquid crystals have been investigated by the ME /100/. Through a careful measurement of the area under the absorption peaks a transition from the solid to the smectic phase has been observed. The area then almost reduces to zero as the sample goes from the smectic to the nematic phase.

Order-disorder transitions in alloys are hard to observe through macroscopic tools. In $FeNi_3$ /38/ such a transition has been detected through the measurement of hf magnetic fields at the Fe site in this alloy. It seems that in studies of order-disorder transition, the ME has considerable advantage.

When certain steels are cooled rapidly a hard material is formed and this is called martensite. The mechanism responsible for transformation is the prototype of a whole class of slid-state transformations called martensitic transformations. Under the tempering action in certain metals or alloys there is a coordinated movement of atoms resulting in a deformation of the lattice and a gain in the strain energy. Some authors prefer to think of such a martensitic transformation in terms of the band Jahn-Teller effect /87/ discussed earlier. Some attempts to establish martensitic transition in REAu alloys using the ^{197}Au resonance have been made by Kimball et al. /29/. A problem with these transitions is that they occur rather sluggishly and show considerable hystersis.

The transformation from crystalline to amorphous state of a material has also been investigated by the ME. In Sb metal, for example, apart from a small CS change across the transition, the Mössbauer resonance fraction has been found to decrease by about 50% in the amorphous state /36/. Such a transformation in Se metal has been probed with the ^{125}Te resonance /37/. The elucidation of the quadrupole patterns in crystalline and amorphous states of Se has led to the understanding of the symmetry of Se chains in the two phases.

Attempts to observe changes in the hf interaction parameters at the superconducting transition have thus far failed /101/; however, possibilities of performing experiments at ultra low temperatures should provide more cases to be investigated.

8. CONCLUSIONS

It was the prime effort of this survey to show on a number of selected examples the contributions the Mössbauer effect has made towards an understanding of the micorscopic phenomena connected with various phase transitions. It has become clear that despite a considerable effort the results are still meagre. This is partly an experimental problem. Better exterimental designs and more efforts toward producing better characterized and purer samples are needed. Hopefully this summary of the already published work will insure more detailed investigations of phase transitions using the Mössbauer effect.

9. ACKNOWLEDGMENTS

This review is based on a discussion held at the conference in which many participants contributed by reporting their work as included here. I wish to thank all of them and in particular A. Biran, L.H. Bennett, R.L. Cohen, J. Danon, I. Deszi, B.D. Dunlap, T. Katila, C. Kimball, H. Montgomery, W.T. Oosterhuis, R.D. Taylor, J.M. Trooster, and

G.K. Wertheim. I am also grateful to B.D. Dunlap, J.M. Friedt, G.M. Kalvius, and D. Schroeer for many helpful discussions and critical readings of the manuscript. I am also indebted to K. Cada for the speed and accuracy with which she processed this manuscript.

REFERENCES

1. H.E. Stanley, INTRODUCTION TO PHASE TRANSITIONS AND CRITICAL PHENOMENA, Oxford: Clarendon Press, 1971; L.P. Kadanoff, W. Götze, D. Hamblen, R. Hecht, E.A.S. Lewis, V.V. Palcianskas, M. Rayl, J. Swift, D. Aspnes, and J. Kane, Rev. Mod. Phys. 39, 395 (1967)

2. M.E. Fischer, Rept. Progr. Phys. 30, Part II, 615 (1967)

3. P. Heller, Rept. Progr. Phys. 30, Part II, 731 (1967)

4. R.S. Preston, S.S. Hanna, and J. Heberle, Phys. Rev. 128, 2207 (1962)

5. D.G. Howard, B.D. Dunalp, and J.G. Dash, Phys. Rev. Letters 15, 628 (1965)

6. D. Gumprecht, P. Steiner, G. Crecelius, and S. Hüfner, Phys. Letters 34A, 79 (1971)

7. H.C. Benski, R.C. Reno, C. Hohenemser, R. Lyons, and C. Abeledo, Phys. Rev. (in press)

8. B.D. Dunlap, M.B. Brodsky, G.M. Kalvius, G.K. Shenoy, and D.J. Lam, J. Appl. Phys. 40, 1495 (1969)

9. G. Groll, Z. Physik 243, 60 (1971)

10. G. Petrich, S. von Molner, and T. Penney, Phys. Rev. Letters 26, 885 (1971)

11. R.L. Cohen, S. Hüfner, and K.W. West, Phys. Letters 28A, 582 (1969); J. Appl. Phys. 40, 1366 (1969); Phys. Rev. 184, 263 (1969)

12. B.D. Dunlap, I. Nowik, and D.J. Lam (to be published)

13. G.K. Wertheim in MÖSSBAUER EFFECT METHODOLOGY, Vol. 4, Ed. I.J. Gruverman, New York: Plenum Press, 1968, p. 159 and reference cited therein.

14. M. Eibschütz, S. Shtrikman , and D. Treves Solid State Comm. 4, 441 (1966)

15. J.M.D. Coey, G.A. Sawatzky, and A.H. Morrish, Phys. Rev. 184, 334 (1969)

16. V.G. Bhide and H.C. Bhasin, Phys. Rev. 159, 586 (1967)

17. J.S. Shier and R.D. Taylor, Phys. Rev. 174, 346 (1968)

18. T. Yamadaya, T. Mitui, T. Okada, N. Shikazono, and Y. Hamaguchi, J. Phys. Soc. Japan, 17, 1897 (1962)

19. M. Tanaka, M. Mizoguchi, and Y. Aiyama, J. Phys. Phys. Soc. Japan 18, 1089,1091 (1963)

20. M. Tanaka, T. Tokoro, and Y. Aiyama, J. Phys. Soc. Japan 21, 268 (1966)

21. V.G. Bhide and M.S. Multani, Phys. Rev. 139, A1983 (1965); 149, 289 (1966)

22. V.G. Bhide and M.S. Hegde, Phys. Rev. B5, 3488 (1972)

23. V.G. Bhide, et al. Solid State Comm. (in press)

24. Y. Hazony, D.E. Earls, and I. Lefkowitz, Phys. Re . 166, 507 (1968)

25. T.J. Gleason and J.C. Walker, Phys. Rev. 188, 893 (1969)

26. M.J. Clauser, Phys. Rev. B1, 357 (1970)

27. P.A. Montano, H. Schechter, and U. Shimony, Phys. Rev. B3, 858 (1971)

28. H. Montgomery (private communication)

29. C.W. Kimball (private communication)

30. D.N. Pipkorn, C.K. Edge, P. Debrunner, G. DePasquali, H.G. Drickamer, and H. Frauenfelder, Phys. Rev. 135, A1604 (1964)

31. I. Dêszi and L. Keszthelyi, Solid State Comm. 4, 511 (1966)

32. L. Asch, J.P. Adloff, J.M. Friedt, and J. Danon, Chem. Phys. Letters 5, 105 (1970)

33. S. Reich and I. Michaeli, J. Chem. Phys. 56, 2350 (1972)

34. G.K. Wertheim, J.P. Remeika, H.J. Guggenheim, and D.N.E. Buchanan, Phys. Rev. Letters 25, 94 (1970)

35. S.L. Ruby (this conference)

36. F.J. Litterst and G.M. Kalvius (to be published)

37. P. Boolchand (to be published)

38. A. Lutts and P.M. Gielen, phys. stat. solidi 41, K81 (1970); A. Heilmann and W. Zinn, Z. Metallk. 58, S113 (1967)

39. G.K. Shenoy, G.M. Kalvius, and S.S. Hafner, J. Appl. Phys. 40, 1314 (1969)

40. G.J. Ehnholm, T.E. Katila, O.V. Lounasmaa, P. Reivari, G.M. Kalvius, and G.K. Shenoy, Z. Physik 235, 289 (1970)

41. T.E. Katila, V.K. Typpi, L. Niinistö, and G.K. Shenoy, Solid State Comm. (to be published)

42. G.J. Long, D. Whitney, and J.E. Kennedy, Inorg. Chem. 10, 1406 (1971)

43. M.H. Villavicencio, P.H. Domingues, and J. Danon (private communication)

44. G.K. Shenoy, G. Abstreiter, and G.M. Kalvius (to be published)

45. P. Heller, Phys. Rev. 146, 403 (1966); P. Heller and G.B. Benedek, Phys. Rev. Letters 8, 428 (1962)

46. R.S. Preston, J. Appl. Phys. 39, 1231 (1968)

47. W.T. Oosterhuis and J.P. Stampfel (private communication)

48. G. Heger and R. Geller, phys. Stat. solidi 53, 227 (1972)

49. G.A. Baker, Jr. Phys. Rev. 124, 768 (1961)

50. C.N. Yang, Phys. Rev. 85, 808 (1952)

51. L.D. Landau and E.M. Lifschitz, STATISTICAL PHYSICS, New York: Pergamon Press, 1958

52. E. Callen and H.B. Callen, J. Appl. Phys. 36, 1140 (1965)

53. V. Jaccarino, L.R. Walker, and G.K. Wertheim, Phys. Rev. Letters 13, 752 (1964); T.E. Cranshaw, C.E. Johnson and M.S. Ridout, Phys. Letters 20, 97 (1966)

54. J. Chappert, Phys. Letters 18, 229 (1965)

55. D. Hone, H. Callen, and L.R. Walker, Phys. Rev. 144, 283 (1966)

56. F. van der Woude and A.J. Dekker, phys. stat. solidi 9, 775 (1965)

57. H.H. Wickman and C.F. Wagner, J. Chem. Phys. 51, 435 (1969); D. Oetridis, A. Simopoulos, and A. Kostikas, Phys. Rev. Letters 27, 1171 (1971)

58. B.I. Halperin and P.C. Hohenberg, Phys. Rev. 177, 952 (1969); K. Kawasaki, Prog. Theor. Phys. Japan 39, 285 (1968)

59. L.M. Levison, M. Luben, and S. Shtrikman, Phys. Rev. 177, 864 (1969)

60. Sh.Sh. Bashkirov and G.Ya. Selyutin, phys. stat. solidi 26, 253 (1968)

61. S. Alexander and D. Treves, Phys. Letters 20, 134 (1966)

62. R. Ingalls, Phys. Rev. 155, 157 (1967)

63. H. Pollak, phys. stat. solidi 2, 417 (1962)

64. R.S. Preston, Phys. Rev. Letters 19, 75 (1967)

65. F.M. Muller, Phys. Rev. 153, 659 (1967)

66. R.S. Preston, I. Goroff, and F.M. Muller, Bull. Am. Phys. Soc. 14, 386 (1969)

67. G.K. Wertheim, D.W.E. Buchanan, and H.G. Guggenheim, Phys. Rev. B2, 1392 (1970)

68. S.S. Nandwani and S.P. Puri, J. Phys. Chem. Solids 33, 973 (1972)

69. T.A. Kovats and J.C. Walker, Phys. Rev. 181, 610 (1969)

70. E.A. Owen and E.W. Evans, Brit. J. Appl. Phys. 18, 611 (1967)

71. Some of the 3d work is given in: G.K. Wertheim, Phys. Rev. 124, 764 (1961); V.G. Bhide and G.K. Shenoy, Phys. Rev. 147, 306 (1966); J.D. Siegwarth, Phys. Rev. 155, 285 (1967); H.N. Ok and J.G. Mullen, Phys. Rev. 168, 563 (1968); M.

Eibschutz, S. Shtrikman, and J. Tenenbaum, Phys. Letters 24A, 563 (1967); G.K. Wertheim, H.J. Guggenheim, H.J. Williams and D.N.E. Buchanan, Phys. Rev. 158, 446 (1967); U. Ganiel, M. Kestigian, and S. Shtrikman, Phys. Letters 24A, 577 (1967); G.R. Hoy and K.P. Singh, Phys. Rev. 172, 514 (1968)

72. W. Zinn and W. Wiedemann, J. Appl. Phys. 39, 839 (1968); R.L. Cohen Phys. Rev. 134, A94 (1964); B.D. Dunlap, M.B. Brodsky, G.M. Kalvius, G.K. Shenoy and D.J. Lam, J. Appl. Phys. 40, 1445 (1969)

73. J.J. Spijkerman, J.C. Travis, D.N. Pipkorn, and C.E. Violet, Phys. Rev. Letters 26, 323 (1971); C.E. Violet and D.N. Pipkorn, J. Appl. Phys. 42, 4339 (1971); T.E. Cranshaw (to be published)

74. See for example, STRUCTURAL PHASE TRANSITIONS AND SOFT MODES, Eds., E.J. Samelsen, E. Andersen, and J. Feder, Oslo: Universitetsforlaget, 1972

75. W. Cochran, Adv. Phys. 9, 387 (1960)

76. G. Shirane and Y. Yamada, Phys. Rev. 177, 858 (1969)

77. H. Unoki and T. Sakado, J. Phys. Soc. Japan 23, 546 (1967)

78. F.W. Lytle, J. Appl. Phys. 35, 2212 (1964)

79. C. Muzikar, V. Janovec, and V. Dvorak, phys. stat. solidi 3, 9 (1963)

80. B.A. Bokov, V.P. Romanov, and V.V. Chekin, Fiz. tverd. Tela 7, 1886 (1965); (sov. phys. solid state 7, 1521 (1965))

81. J.M. Trooster, phys. stat. solidi 32, 179 (1969)

82. R.R. Hewitt, Phys. Rev. 121, 45 (1961)

83. V. Dvorak, phys. stat. solidi 14, K161 (1966)

84. For a detailed account see for example, R. Englman, THE JAHN-TELLER EFFECT IN MOLECULES AND CRYSTALS, New York: Wiley-Interscience, 1971

85. G.L. Bacchella, P. Imbert, P. Meriel, E. Martel, and M. Pinot, Bull. Soc. Sci. Bretagne 39, 121 (1964)

86. F. Hartmann-Boutron, J. Phys. Rad.,29, 47 (1968)

87. J. Labbé and J. Friedel, J. Phys. Rad. 27, 153 (1966); 27 303 (1966); 27 708 (1966)

88. A.R. Bates and K.W.H. Stevens, J. Phys. C2, 1573 (1969)

89. C. Trapp and C.I. Shyr, J. Chem. Phys. 54, 196 (1971)

90. H. Van Kempen, W.I. Duffy, A.R. Miedema, and W.J. Huiskamp, Physica 30, 1131 (1964)

91. L. Asch and J.M. Friedt, PROCEEDINGS OF THE CONFERENCE ON MÖSSBAUER SPECTROMETRY, Dresden, DDR: Physikalische Gesellschaft, p. 447

92. For a detailed review see for example, D. Adler, Rev. Mod. Phys. 40, 714 (1968)

93. J.C. Slater, Phys. Rev. 82, 538 (1951)

94. N.F. Mott, Phil. Mag. 6, 287 (1961)

95. T. Shinjo and K. Kosuge, J. Phys. Soc. Japan 21, 2622 (1966)

96. D.B. McWhan, T.M. Rice, and J.P. Remeika, Phys. Rev. Letters 23, 1384 (1969)

97. G. Shirane, D.E. Cox, and S.L. Ruby, Phys. Rev. 125, 1158 (1962)

98. J.M. Friedt (private communication)

99. R. Ingalls, H.G. Drickamer, and G. DePasquali, Phys. Rev. 155, 165 (1967)

100. D.L. Ulrich, J.M. Wilson, and W.A. Resch, Phys. Rev. Letters 24, 355 (1970)

101. V. Vali, T.W. Nybakken, and J.G. Dash, Rev. Mod. Phys. 36, 359 (1964)

THE STUDY OF VIBRATIONS IN CRYSTALS USING
THE RAYLEIGH SCATTERING OF RESONANT GAMMA RAYS

D.A. O'Connor

Department of Physics, University of Birmingham

Birmingham, B15 2TT, England

1. INTRODUCTION

An essential feature of the Mössbauer effect is that the recoilless gamma rays have wavelengths in the same range, about 0.1nm, as the X rays used in the X-ray diffraction study of crystals. The Debye-Waller factor for X rays, $\exp - \frac{1}{2}\langle((\underline{K}_1-\underline{K}_0)\cdot\underline{u})^2\rangle$, becomes the recoilless fraction in the Mössbauer case when the net momentum transfer $\underline{K}_1-\underline{K}_0$ for the X rays is replaced by the recoil momentum $\underline{K}$ for the gamma ray. The recoilless fraction of X-ray scattering is identified with the Laue spots and Bragg peaks of the X-ray diffraction pattern and the X-ray crystallographer uses resolution in momentum space to separate out the elastic part of the scattering. This separation is imperfect but can be radically improved if the X rays are replaced by Mössbauer gamma rays and the energy resolution of the Mössbauer experiment is combined with the momentum resolution of the X-ray experiment. The momentum resolution of this combination can be made to be of the same order as in X ray or neutron diffraction but the energy resolution, typically 10^{-8}eV will be much better than for the neutron experiment, typically 10^{-4} eV. The high energy resolution of the Mössbauer experiment in fact practically prevents its use in the observation of single phonon lines, as in the neutron inelastic scattering experiments, since phonon lines are too broad, $\sim 10^{-4}$eV. However, as we shall show below, there are situations in which the high energy resolution can be put to good use.

The first use of the Mössbauer effect to distinguish between the elastic and inelastic Rayleigh scattering was by Tzara and Barloutard (1960) (1) who used polycrystalline scatterers. Essentially the same technique was applied by O'Connor and Butt (1963) (2) to the scattering by single crystals.

2. RECOILLESS FRACTION MEASUREMENT

Consider a scattering experiment in which the radiation from a single line source of recoilless fraction f_s is detected after scattering through an angle 2θ by a non-resonant scatterer which scatters a fraction f_c without recoil. If all the inelastic scattering processes are considered simply to remove radiation from the narrow line and not broaden it, then in the scattered beam a fraction $f_s f_c$ is still resonant. A single line absorber, matched to the source, is placed between the scatterer and the detector. If $I_\theta(0)$ and $I_\theta(\infty)$ are the background-corrected counting rates with the absorber at rest and vibrating respectively then the fractional resonant effect will be

$$R(\theta) = \frac{I_\theta(\infty) - I_\theta(0)}{I_\theta(\infty)} = f_s f_c (1 - x)$$

where x is the fraction of resonant radiation transmitted by the absorber at rest. The factor $f_s(1 - x)$ can be determined by measuring the fractional resonant effect, $R(0)$, of the incident beam either by removing the scatterer and placing the detector in the incident beam or by placing the absorber between the source and the scatterer and repeating the rest and motion measurements. The second method is to be preferred since the radiation received by the detector is essentially the same as in the measurement with the absorber following the scatterer and counting rates, backgrounds etc. are similar. Thus the recoilless fraction of the scatterer f_c can be found from the ratio $f_c = R(\theta)/R(0)$. The total radiation scattered $I_\theta(\infty)$ can then be divided into an elastic component $I_e(\theta) = f_c I_\theta(\infty)$, and an inelastic component $I_i(\theta) = (1 - f_c) I_\theta(\infty)$.

This classification into elastic and inelastic components uses the source-absorber line width as a borderline. The inelastic scattering in a crystalline specimen will be due to the Compton effect and phonons, both of which processes produce energy changes large compared with the effective line width of 10^{-8}eV, say for Fe57. The proportion of phonons with energy 10^{-8}eV is negligibly small so that line-broadening is not expected and the elastic/inelastic classification is not arbitrary but for practical purposes exact.

The elastically scattered radiation for an ordered crystalline material will be concentrated in diffraction peaks and its intensity will depend on geometry, crystal texture etc. The Compton component of the inelastic scattering has a predictable smooth variation with angle but the phonon scattering will be significantly peaked with angle. Thus the fraction f_c is highly geometry-dependent and not easily interpretable. If the angular resolution of the experiment is deliberately made poor so that the diffraction peaks are averaged, and a calculated correction for the Compton scattering is made, then the ratio of elastic to (total-Compton) can be set equal to the Debye-Waller factor $\exp - \frac{1}{2} \langle ((\underline{K}_1 - \underline{K}_0) \cdot \underline{u})^2 \rangle$. This is in effect the method used by Tzara and Barloutard (1) and in modified form for amorphous scatterers by Kroy and Vonach (3). An important point to note here is that since the momentum transfer $\underline{K}_1 - \underline{K}_0$ depends on angle it can be made smaller by the factor 2sinθ than the momentum transfer in the straightforward Mössbauer effect so that larger values of $\langle u^2 \rangle$ should be measureable by this method and of course it can be applied to substances which do not have a Mössbauer transition.

The accuracy of recoilless fraction measurements depends on the choice of source and absorber and a few general remarks should be made here about experimental details. The observed resonant effect is $I_0(\infty) f_s f_c (1-x)$ so that for an accurate measurement of f_c both $I_0(\infty)$ and $f_s(1-x)$ should be as large as possible. This means that the source should be of high activity with a large recoilless fraction f_s and the absorber should be as black as possible. We use 100mCi Co57 in Rh sources with $f_s \sim 0.7$ and the fluoferrate black absorber. The gamma ray wavelength, 0.086nm, is comparable with X-ray wavelengths and gives conveniently large Bragg angles of scatter. For work on single crystals narrow beams of radiation are required and we use a thin foil source of a few millimetres diameter mounted in heavy metal shielding to produce four beams emerging from both faces of the source at angles of 45° to the plane of the foil.

In work on single crystals the crystal itself, set at a Bragg angle, will select the gamma ray so that the detector need not have good energy resolution. On the other hand a weak reflection from a small crystal may produce counting rates of a few counts per second so that the background counting rate should be as small as possible. This can be achieved with NaI scintillators of small volume or, better, with a silicon detector.

3. SCATTERING BY SINGLE CRYSTALS

3.1 General Principles

The interpretation of the elastic fraction of scattering becomes particularly simple if use is made of single crystal specimens which will concentrate the elastic scattering in Bragg peaks. The general features of this situation are illustrated in Figure 1. The Bragg elastic peak is superimposed on a peaked phonon-scattered component and a Compton background. The angular width of the elastic peak depends on the angular resolution of the apparatus and the crystal mosaic spread and is independent of Bragg angle.

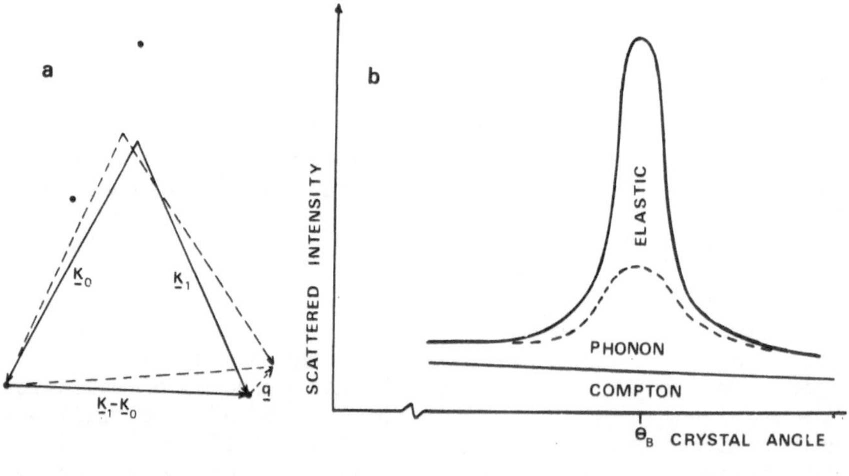

Figure 1(a) The scattering process in reciprocal space, elastic-full line, phonon-dashed line.

(b) Scattered intensity versus angle at the Bragg position.

The peak height can therefore be taken as proportional to the integrated intensity of the reflection and we can write

$$I_e(\theta) \propto \left| \sum_i F_i(\theta) e^{-W_i} \right|^2 \qquad 3.1$$

where $F_i(\theta)$ is the structure factor of the i-th atom in the unit cell and W_i the corresponding Debye-Waller factor, $W_i = \frac{1}{4} \langle ((\underline{K}_1 - \underline{K}_0) \cdot \underline{u}_i)^2 \rangle$. Measurements of $I_e(\theta)$ as a function of temperature can give the variation of $\langle (u_i)^2 \rangle$ with temperature and a measure of the characteristic temperature $\Theta(-2)$.

The peaked part of the phonon scattering is principally due to one-phonon scattering which varies approximately as $(\Delta\theta)^{-2}$ where $\Delta\theta$ is the deviation from the Bragg angle. The intensity of the phonon scattering relative to the elastic scattering depends on the angular resolution, it increases with increasing acceptance angle of the detector, but in any case is proportional to

$$\left| \sum_i F_i(\theta) e^{-W_i} (\underline{K}_1 - \underline{K}_0) \cdot \underline{u}_i \right|^2 \qquad 3.2$$

The difference between $I_t(\theta)$ at the peak and at the flat wings of the $I(\theta)$ curve can be taken as approximately proportional to the one-phonon scattering. For a monatomic lattice the structure factor and Debye-Waller factor cancel in the ratio of inelastic to elastic scattering to give a measure of $((\underline{K}_1 - \underline{K}_0) \cdot \underline{u})^2$.

A final general point to note is that the inelastic phonon scattering observed at the Bragg peak is due to low-wave-vector phonons so that changes in the inelastic scattering observed in an experiment can be identified as due to phonons in a range of wave vector defined by the angular resolution. Changes in the Debye-Waller factor which are not accompanied by a change in the inelastic scattering must be due to phonons outside this range.

3.2 Measurements of Debye Temperatures

The determination of Debye-Waller factors for X-ray reflections depends on the measurement of the integrated intensity of diffraction peaks which are assumed to be due to elastic scattering. A background correction determined from the flat wings of the curve is subtracted from the peak to account for inelastic scattering. Referring to Figure 1 it is seen that this method systematically includes the peaked part of the inelastic scattering and since this has the same dependance on $\langle u^2 \rangle$ 3.2, as the Debye-Waller factor 3.1, this procedure leads to a systematic underestimate of $\langle u^2 \rangle$ and therefore an overestimate of the Debye temperature. Butt and O'Connor (1967) (4) found that for Al and KCl the Debye temperature calculated using the elastic scattering a few percent lower than that calculated by the X-ray method using the total scattering for reflections in which the inelastic scattering amounted to up to 20% of the total.

For high angle reflections and high temperatures the systematic error becomes more significant. Ghezzi et al (1969) (5) have shown that the inelastic scattering for the (444) and (555) reflections in perfect silicon crystals can amount to 50% of the total in typical geometry. Similar proportions obtain in crystals of NaCl and KCl at temperatures above about 200C where $\langle u^2 \rangle$ begins to show a non-linear dependance on temperature. Mössbauer scattering measurements in these cases have made possible the determination of the quartic term in the Debye-Waller factor

$$\exp\left\{-\tfrac{1}{2}\langle(\underline{K}\cdot\underline{u})^2\rangle + \tfrac{1}{24}\left[\langle(K\cdot u)^4\rangle - 3\langle(K\cdot u)^2\rangle^2\right]\right\}$$

due to anharmonic effects for NaCl (6) and KCl (7).

3.3 Anomalous Vibration in Aluminium

In the course of work on aluminium single crystals we have noticed an anomalous behaviour in the Bragg scattering in the region of 180C (8). The total, elastic and inelastic scattering versus temperature curves are smooth for a well-annealed but for a freshly-cut and etched crystal slice there appears a dip in the total and elastic scattering and a peak in the inelastic scattering at temperatures just below 180C, see Figure 2. The anomaly is absent when the curve is retraced slowly but can be made to reappear if the crystal is heated and quenched. It seems reasonable therefore to associate the anomaly with the presence of defects. The increase in inelastic scattering is due to an increase in the excitation of long-wavelength phonons presumably associated with the motion of defects. X-ray measurements might have detected the anomalous dip in the total scattering which might have been interpreted as due to a change in

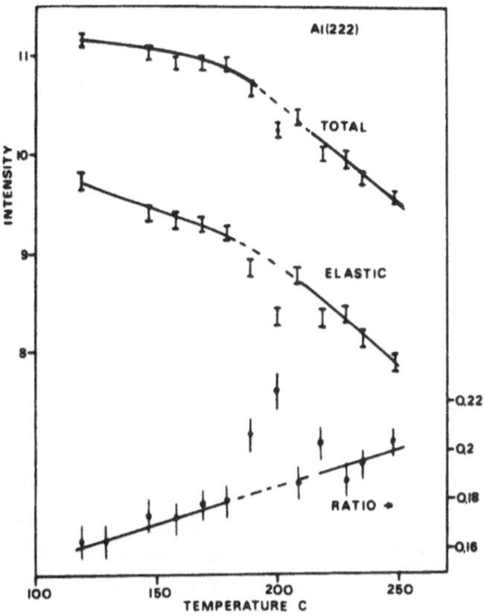

Figure 2

crystal texture, but the Mössbauer measurements alone could reveal the large increase in inelastic scattering which must imply an increased thermal motion of the atoms in the crystal.

3.4 The Acousto-electric Effect in CdS.

This is another experiment in which related changes in both the elastic and inelastic scattering were observed (9). When the drift velocity of conduction electrons exceeds the sound wave velocity in CdS energy is transferred from the electrons to the sound waves and sound waves can be amplified, see for example White (1962) (10). This effect is usually demonstrated by injecting ultrasound but in this experiment the existing thermal vibrations were amplified. Mössbauer gamma rays were Bragg-scattered from the (0002) planes of the crystal and a current passed parallel to the reflecting surface. The electrons are coupled to transverse sound waves so that any excess vibration would produce a change in the elastic scattering. The observed change in the elastic scattering was used to calculate, through the Debye-Waller factor, the excess vibration due to the amplification of the thermal vibrations. The excess amplitude, up to 0.04 nm, increased from the threshold voltage at which the non-ohmic current-voltage characteristic showed that energy was being transferred from the electrons to the phonons. The total scattering in the Bragg peak did not depend on the current through the crystal showing that the excess phonons produced were of long wavelength and that the extra inelastic scattering they produced exactly compensated the reduction in the elastic scattering. An X-ray experiment detected no change at all.

3.5 Soft Modes in $BaTiO_3$

The transition in $BaTiO_3$ from the paraelectric cubic phase to the ferroelectric tetragonal phase has been explained (11) as being due to the progressive lowering in frequency, or 'softening', of a zero wave vector optic mode, as the temperature is lowered towards the Curie temperature, 120C. The lowering in frequency results in an increased vibrational amplitude in this mode which should produce an increase in the inelastic scattering at Bragg peaks since it is of zero wave vector. Mössbauer scattering measurements (12) have indeed shown that the elastic scattering falls below the value expected from the higher temperature behaviour as the temperature is reduced towards the Curie temperature and the inelastic/elastic ratio shows evidence of the extra inelastic scattering, see Figure 3a. The excess vibrational amplitude calculated from the Debye-Waller factor was found to be comparable with the relative distortion between cubic and tetragonal phases. Similar behaviour is observed (13) at the tetragonal-orthorhombic transition at 5C where there is a very marked increase in the inelastic scattering as the

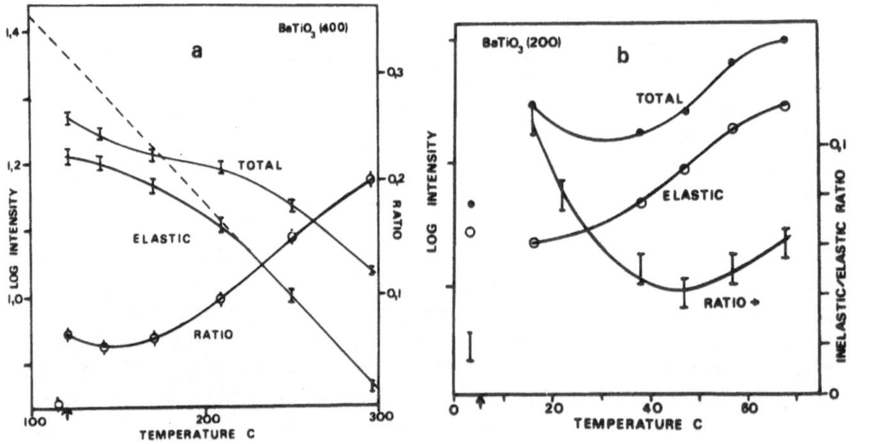

Figure 3. Scattering from $BaTiO_3$. Transition temperatures indicated by arrows.

temperature is lowered, Figure 3b, suggesting that a soft mode is also responsible for this transition.

3.6 Excitation of Vibrations in Cubic $BaTiO_3$ by an Electric Field.

The cubic phase of $BaTiO_3$ is characterised by an extremely high dielectric constant so that application of an oscillating electric field might produce a measurable ionic vibration. Such a vibration would be of the optic type with ions of opposite sign vibrating in antiphase, and of a zero wave-vector so that the scattering from such a mode would occur at a Bragg peak and be distinguishable by the Mossbauer scattering method if the applied field were of sufficiently high frequency. The amplitude of the coherent wave scattered by such a crystal is proportional to

$$\sum_i F_i(\theta)\, e^{-W_i} \sum_p J_p(Ku_i)\, e^{ip(\omega t + \phi_i)}$$

where $J_p(Ku_i)$ is the pth Bessel function, u_i the amplitude of the ith atom at the driving frequency ω with relative phase ϕ_i. The intensity contains a central peak depending on the J_0's and sidebands J_1, J_2 etc. of frequency shift $\pm\omega$ $\pm2\omega$ etc.

In experiment (14) an RF current of up to 0.5A at 10V at a frequency of 13MHz was passed through a $BaTiO_3$ crystal 0.2mm thick at temperatures above the Curie temperature. The change in the elastically scattered radiation and the relative size of the first sideband were measured by comparing the absorption of the scattered beam in a black absorber and in an enriched stainless steel absorber which overlapped the sidebands to different extents. Values of $(\underline{K}.\underline{U})$ for a simple monatomic model were deduced from the measurements for the (200) (400) and (600) reflections for different temperatures and RF voltages, see Figure 4. The curve of $(\underline{K}.\underline{U})$ versus v shows a saturation effect and whereas the saturation value for (600) is roughly 3/2 times that for (400) the (200) value is in proportion too large. From this we deduce that the vibration amplitude is larger at the surface. The actual saturation amplitude measured, of the order of 0.01nm is an order of magnitude greater than that calculated from the current and the dielectric constant, assuming two charges per ion. Finally the variation of amplitude with temperature is quite unlike the variation of dielectric constant. The velocity spectrum for the (200) and (400) spectra revealed the presence of both the first and second sidebands, and the ratio

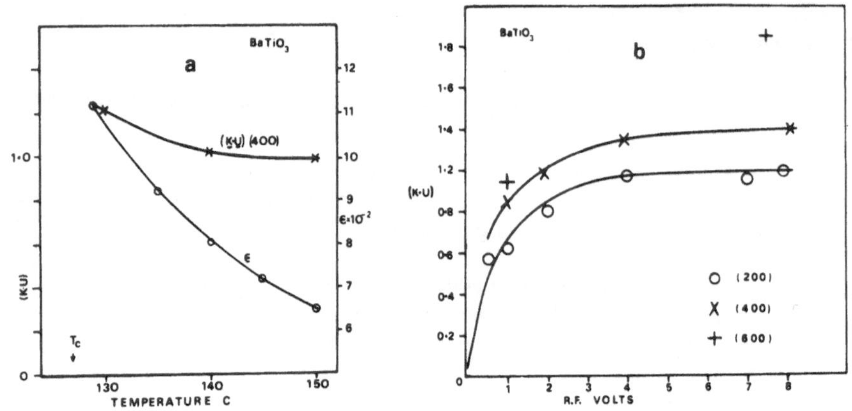

Figure 4. Amplitude versus (a) temperature, (b) R.F. voltage

of second to first sideband intensity was 50% greater than would be expected if the atoms were all moving in phase. The first sideband intensity depends on terms linear in the displacements but the second sideband depends on quadratic terms so that an antiphase motion of for example Ba^{++} and TiO_3^{--} would explain this result. We conclude from all this that in the high temperature cubic phase there are piezoelectric surface defects. This would explain the antiphase motion, the saturation with voltage and the variation of amplitude with depth. Such defects have been inferred in other experiments (15).

References

1. Tzara,C. and Barloutard,R.,(1960),Phys.Rev.Letters 4,405.
2. O'Connor,D.A. and Butt,N.M.,(1963),Phys.Letters 7, 233.
3. Kroy,W. and Vonach,H. (1969), Z. Angew. Phys. 27, 335.
4. Butt,N.M. and O'Connor,D.A.(1967),Proc.Phys.Soc. 90,247.
5. Ghezzi,C.,Merlini,A. and Pace,S.(1969),Nuovo Cimento 64B103.
6. Butt,N.M., and Solt,G. (1971) Acta Cryst. A27, 238.
7. Solt,G.,Butt,N.M. and O'Connor,D.A.(1972) to be published.
8. Martin,C.J. and O'Connor,D.A. (1972) to be published.
9. Horley,C.C.Lomer,T.R.O'Connor,D.A.(1969)J.Phys. C. 3,L7.
10. White,D.L. (1962) J. Appl. Phys. 33, 2547.
11. Cochran,W. (1960) Advances in Physics 9, 387.
12. O'Connor,D.A. and Spicer,E.R.(1969)Phys.Letters 29A,136.
13. Fitzgerald,W.J.(1972)M.Sc.Thesis,University of Birmingham.
14. Kilner,J.A.and O'Connor,D.A.(1972) to be published.
15. Chynoweth, A.G. Phys. Rev. 102, 705.

MÖSSBAUER STUDIES OF AQUEOUS LIQUIDS AND GLASSES*

S. L. Ruby

Argonne National Laboratory

Argonne, Illinois 60439

ABSTRACT

A brief review of the classic distinctions between glasses, liquids, and crystals is given. Using ferrous chloride in H_2O as a model system, how these distinctions appear in Mössbauer spectroscopy is shown. The shape of the narrow absorption lines produced in these glasses is discussed, and argument is made for nearly perfect short-range order. The second system discussed is ferrous ion dilute in $H_3PO_4 \cdot H_2O$. This solution is most difficult to crystallize, allowing long measurements well above the glass-transition temperature T_g. The main effects with increasing temperature in this system are decreasing amplitudes and broadening line widths. In addition we see here clear changes in the slope vs. temperature at T_g of such parameters as line width, quadrupole splitting, and area. The quadrupole splitting decreases more rapidly with temperature above T_g than suggested by extrapolation from lower temperatures; a relaxation of the direction of the EFG, as calculated by Blume, seems to be the explanation.

*Work performed under the auspices of the U. S. Atomic Energy Commission.

INTRODUCTION

During the last several years as I learned about this subject, various friends and co-workers were my main teachers. I'd particularly like to mention Hwa-Cheng Chang, John Stevens, Bruce Zabransky, Shmuel Bukshpan, Israel Pelah, and Paul Flinn. Also, one has in this field a large group of allies publishing similar work. Although because of time pressure in this talk I shall restrict myself to showing spectra and tables from our work, it should be understood that much of what you hear today is the collective result of a rather large group of workers. A few of their names include Harry Bernas who early produced a wise paper[1] which has had less influence on further progress than its insight invited. Lajos Keszthelyi and Istvan Dezsi in Budapest initiated systematic studies involving several iron-anion combinations, and have continued to contribute the largest volume of new knowledge.[2] Morton Kaplan and Arthur Nozik followed up that work both experimentally and theoretically.[3] In organic liquids, Paul Craig's pioneering work in glycerol[4] was followed by results from Champeney, Woodham,[5] Elliot, Hall and Bunbury,[6] and now Mullen.[7] R. Cohen,[8] T. Sonnino,[9] S. Bukshpan,[10] and Simopolus and Wickman[11] should be mentioned as well as J. H. Jensen's[12] exciting theoretical and experimental results in propane.

The plan of this paper is to review the classic distinctions between glass, liquid and crystal and to illustrate how the Mössbauer effect adds to such traditional measurements as specific heat, differential thermal analysis, and specific volume. Our model here system is ferrous chloride in water. The narrow lines observed in the glass are discussed and explained as nearly perfect short-range order, a rather important result. Another system, dilute ferrous ion in concentrated phosphoric acid ($H_3PO_4 + H_2O$) which is easier to keep in the liquid state, is employed to show the changes in slope of such quantitites as $x^2(T)$, $\Gamma(T)$, QS(T), etc. at T_g. The quadrupole splitting decreases faster with increasing temperature than expected and the result is consistent with relaxation of the crystal field axes, as calculated by Blume.[13]

WHAT IS THE MICROSTRUCTURE OF A GLASS?

There is a textbook figure which has been around for over thirty years which is expanded upon in Fig. 1. In a, b, and c, the specific heat, the specific volume, and the

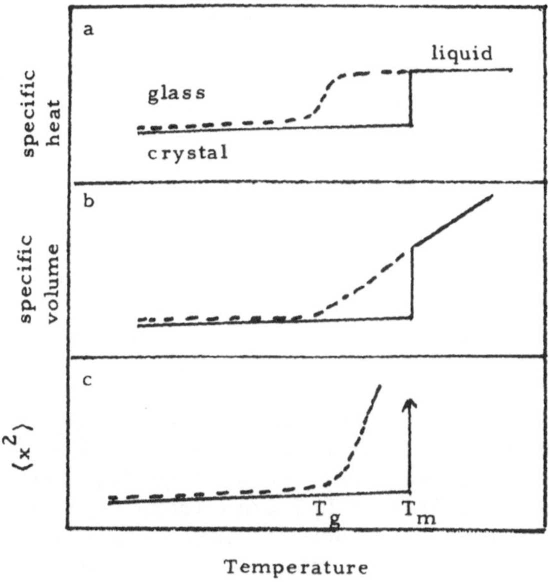

Figure 1

squared amplitude of atomic vibration are plotted vs. temperature for some simple material. The melting temperature Tm separates the liquid and crystalline regions. However, in some cases, crystals do not always form as the liquid is cooled. There is another interesting and subtle subject called nucleation which we will not discuss here, except to say that it can provide a fertile field for applications of Mössbauer spectroscopy. If no crystallization, the state can move along the dotted curves, through the super-cooled liquids and into the glassy state. T_g indicates the glass transition temperature which is not uniquely defined; it extends over a range of several degrees.

A phenomenological description of these effects is that as the warm liquid cools, it is still an equilibrium fluid, characteristic relaxation times are tiny parts of a second, and both atomic rearrangements and atomic vibrations contribute to the specific heat and volume. On the other hand, when the liquid is very cold (glass), the relaxation times are many centuries, and atomic rearrangements are no longer seen. The glass transition is the point where the relaxation times go from seconds to days. In this view, there is no qualitative difference between glass and liquid, but merely that the glass at T has the frozen-in structure of the liquid at T_g. What is surprising is how abrupt, how nearly first-order-like, is T_g. There is no accepted theory of this transition now. A so-called "free-volume" model does explain some features fairly easily.[14] The Mössbauer studies provide new information not only on $\langle x^2 \rangle$ as seen in 1(c), but also on jumping times and/or diffusion constants as seen from line broadening.

In Fig. 2 temperature and absorption amplitude are plotted vs. time for a tin solution. The example is being heated and cooled linearly in time, and the notches on the temperature curve show exothermic (crystallization) and endothermic (melting) points. The purpose is to remind you of the kinds of confusion in the early experiments when poorly understood solutions were being glassed by quick-freezing, and warmed to super-cooled liquids, then crystallizing to several different hydrates (some days of the week) and in general rather too much new data with almost every slight variation of the experiments. The main points are the rapid decrease in effect in the liquid and its abrupt reappearance coincident with crystallization. In Fig. 3 a differential thermal record for a 30 w/o $FeCl_2$ + 18 H_2O) solution is shown on the left. The glass transition, melt-

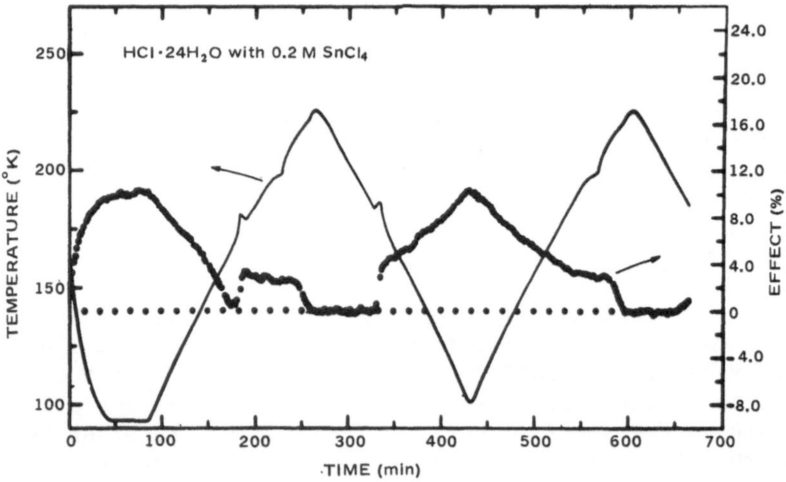

Figure 2

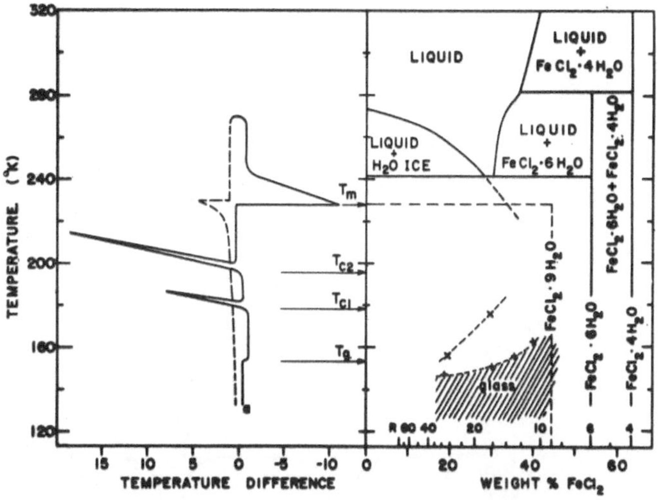

Figure 3

ing, and <u>two</u> large exothermic crystallizations are indicated. By repeating such measurements vs. concentration, plus a good deal of MB studies to make sure when the crystals were or were not present, the phase diagram in the right was constructed. It does not give much insight into such kinetic complexities as why $FeCl_2 \cdot 6H_2O$ does not crystallize from the melt even after several days at 235°K. For our purposes today, we restrict ourselves to the above-mentioned concentration and discuss three special states.

1) $FeCl_2 + 18H_2O$ as a glass between T_g and T_{c1}

2) $FeCl_2 + 9H_2O$ as a glass between T_{c1} and T_{c2}

3) $FeCl_2 \cdot 9H_2O$ crystals between T_{c2} and T_m.

In Fig. 4 we see the spectra of these three materials. Here each has been fit by a simple pair of Lorentzians. (1) and (2) are nearly identical! Here $\Gamma \cong 0.5$ mm/sec with QS $\cong 3.35$ mm/sec, but even visual inspection shows the fit is poor (Misfit = $(0.5 \pm .005)\%$).[15] On the other hand, (3) has $\Gamma = 0.3$ mm/sec and QS = 3.73 mm/sec with Misfit down to $(0.06 \pm .01)\%$. This represents a good fit for so few parameters. Physically, (3) represents a simple crystalline

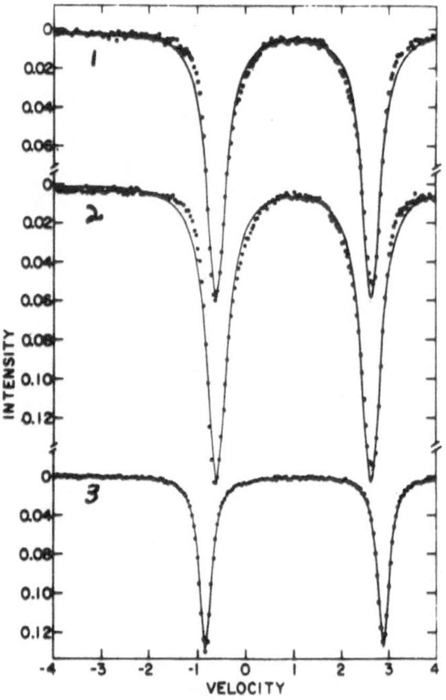

Figure 4

powder. It is pertinent to compare (3) with $FeBr_2 \cdot 9H_2O$ spectra - in contrast with $FeCl_2 \cdot H_2O$, these crystals can be prepared conventionally. In each of the crystals, each iron ion is surrounded by six waters. This comparison has been just made by Dezsi[2] - the two spectra are indistinguishable. This is clear proof that the kind of anion in the second nearest neighbor positions has little effect on the ferrous hexaquo-ion here under study. Further comparison with hydrates of ferrous sulfate and of ferrous fluorosilicate containing the same iron ion lead to the conclusion that this ion gives a normal Ingall's-type ferrous splitting of about 3.5 mm/sec whenever surrounded by 6 water ions, regardless of other ions outside the nearest neighbor shell. The magnitude of the measured splitting does vary modestly from one crystal to another, mainly due to different $q_{lattice}$.

However, if we change the nearest neighbors of the ferrous ion, then the spectra are changed more drastically.

Table I

crystals, n =	0	2	4	6	9
quadrupole splitting (mm/sec)	1.10	2.59	3.08	1.71	3.73

If we consider the series of crystals ($FeCl_2 \cdot nH_2O$, n = 0, 2, 4, 6, 9), only the last of these contains the hexaquo-ferrous ion. The others have local structures of the type $Fe(H_2O)_{6-x}\,Cl_x$. In Table I, we show the quadrupole splitting associated with each crystal. It is evident that the spectra 1 and 2 obtained from the glasses contain very little, if any, of such components.

Numerous attempts have been made to fit the glassy spectra well. One approach has been to assume that the glass consists of m micro-environments, each characterized by quadrupole split doublet. Of course, by making m large enough, one can make the fits better and better. However, to get good fits, we need $m \geq 4$, large line widths, and the values for the various QS do not repeat any of those in the table. The glass spectra described as the sum of a few micro-environments just doesn't seem to fit the data.

On the other hand, if we go to a distribution of quadrupole splittings, we could always arrange an excellent fit. What is pleasing to us is that a very simple, one parameter, distribution can (nearly) do the job. We replace a delta function distribution in e^2qQ () by an "exponential" distribution () with $N(v) = (\beta/2)\,(\exp[-B(v_+-v)] + \exp[-\beta(v-v_-)])$. The addition of the free parameter β to the calculated spectrum improves Misfit to $(.08 \pm .01)\%$. The value found for $\beta \cong 4.5$. At this stage, our interpretation is that any distribution that allows "in-board" blurring of about $1/\beta = 0.25$ mm/sec will fit the glassy spectra well.

It appears quite likely that every ferrous ion in these aqueous glasses is surrounded by six waters. At a particular site there are extra $q_{lattice}$ fields and these typically lower the observed field since $q_{lattice}$ and q_{ion} are expected to be of opposite sign.[3] Since the type and positions of second (and further) nearest neighbors vary, all iron ions do not feel the same $q_{lattice}$ and thus an "exponential" fit could arise from the superposition of a range of $q_{lattice}$. This is only plausibility and not a proof, but it is strongly felt here that nearly perfect short-range order has been demonstrated in these glasses. It would appear very difficult to maintain any microcrystalline model after this. It is perhaps worthy of mention that the continuous random network model suitable for single component systems[16] could also have explained our spectra, but it is very difficult to see how to use that model for varying concentrations of water.

WHAT CAN BE LEARNED ABOUT ATOMIC MOTIONS IN SCL?

Since P. Craig's early work on glycerol, iron ions have been used as probes of atomic motion in a variety of systems, including liquids based on methanol and propane. The work reported here involves another liquid-concentrated phosphoric acid. The mixture is $H_2PO_4 + H_2O$; if this liquid is diluted more than about a factor of two, then crystallization of nearly pure water ice upon freezing returns the molecular ratio in the glass to one. Iron metal placed in it dissolves readily to the ferrous ion, and by using enriched metal, the molar abundance of ferrous ion was kept below 0.003. The spectra are very similar in appearance to those seen in the ferrous chloride glasses, and the iron atoms must again be surrounded by six waters. However, the phosphate ions manage to form a bridged structure, somewhat similar to the linking of SiO_2 molecules in quartz, which is very effective at preventing crystallization.

We have made DTA measurements on this liquid, the glass transition temperature T_g was found at about 175°K. In addition, we have examined the specific volume, as shown in Fig. 5. Fitting a straight line to the low and high temperature regions, T_g = 174.5°K. After this a long series of Mössbauer measurements were made with $78°K < T < 250°K$. Both narrow and wide velocity scans were employed. The individual spectra reported here were fitted with a simple symmetric quadrupole doublet; the parameters were amplitude, width , isomer shift S, and quadrupole splitting QS. The spectra are not well fitted by this procedure, and the fittings are being repeated using the "exponential" distribution.

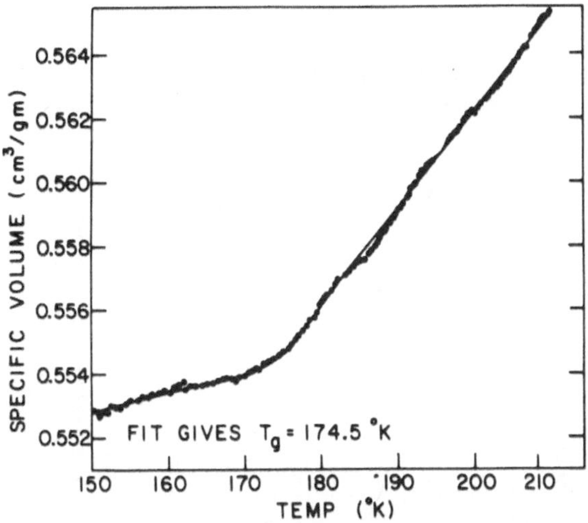

Figure 5

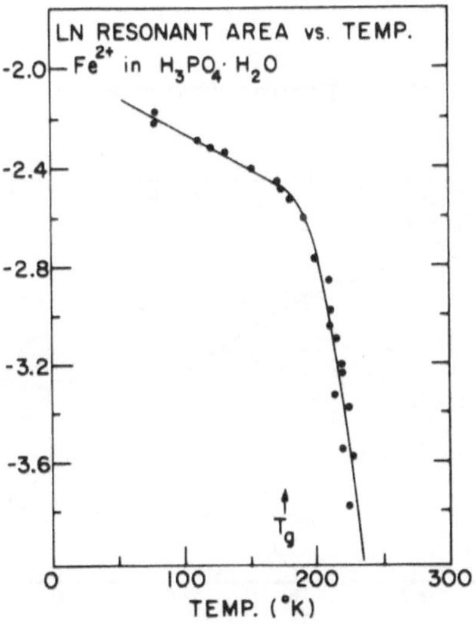

Figure 6

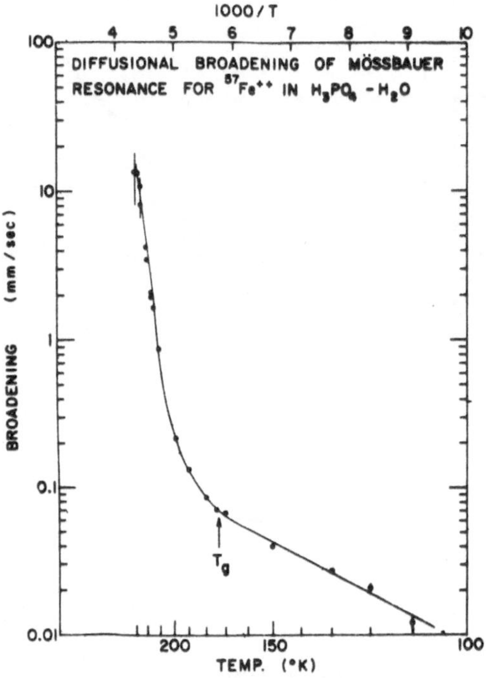

Figure 7

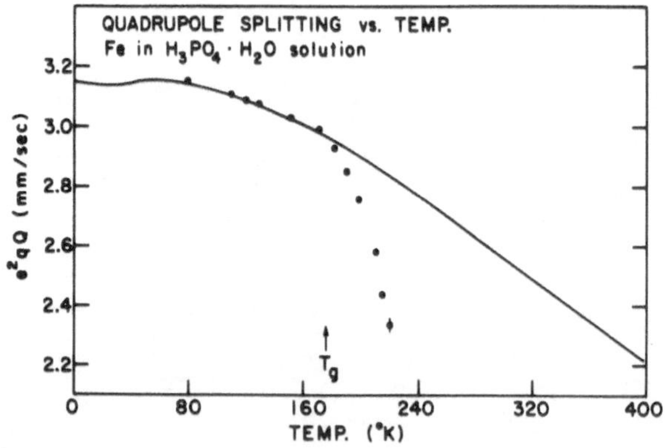

Figure 8

However, even with simple fits, values can be obtained for area, $\Gamma - \Gamma_0$ where Γ_0 is the experimental width at 78°K, and QS. These results are displayed in Figs. 6 - 8. It is interesting to see how clearly T_g appears in the Mössbauer parameters. A detailed analysis of these spectra is nearing completion, even though it is a matter of some delicacy. Numerous physical effects are operating here simultaneously and it is easy to become adrift with too many free parameters on what are, individually, rather featureless spectra.

As an example, consider line broadening. This is affected by a) the distribution of $q_{lattice}$ in the various micro environments of the glass, b) thick absorber "blackness" effects at the lower temperatures, c) normal crystalline-type jump diffusion, most simply seen below T_g, d) a relaxation of the direction of the axes of the EFG at each site, leading to another parameter which can be called the rotation time (1/w).

To have any hope of separating these effects, one must use the whole data more as one group. For example, c) and d) above both broaden the lines, but d) also "collapses" the QS toward zero splitting as w increases. This theory is due to M. Blume.[14] In Fig. 9 we show some results of his analysis of the spectrum shape for various w. As one would anticipate, for very rapid rotation, the special axis is quite lost and the spectrum collapses to a singlet. In Fig. 8 we

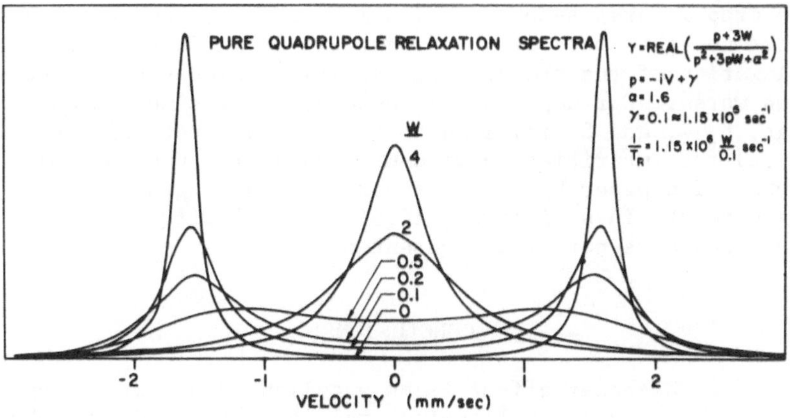

Figure 9

have fitted the low temperature QS with an Ingall's type temperature dependence (the solid line). The rapid drop of QS below this above Tg is at least partly due to the Blume mechanism. One must remember that ordinary jump diffusion merely widens lines, but does not shift them.

Since the analysis is not complete, I can conclude with a word about what one hopes to get (one important attribute for physicists is hopefulness). Atomic motions in liquids has been an important area of research for a long time, and a wide variety of experimental and theoretical tools have been utilized. The theorist would most like the experimentalist to measure G(r,t), the space-time auto-correlation function. For solid-state physics, most of the interesting things about G have become asymptotic by 10^{-11} sec. Mössbauer studies, which are sensitive to much longer times, have been useful only in measuring the mean time between diffusion jumps to new lattice positions, τ_J. For viscous liquids, G(r,t) apparently has structure out to much longer times than for solids. Some NMR people talk of "ultraslow" motions in liquids. Sometimes they may be relatively simple like rotation rates for large molecules.

In any case, in addition to line broadening, our resonance measurements get a grip on the area, proportional to f. The simple jump model does not predict much of a decrease in f with higher temperature, but mainly broader and broader lines. If all the broadening seen in Fig. 7 is attributed to more and more rapid "jump" diffusion, then the drop in area seen in Fig. 6 is not understandable. It seems now that much of the broadening is in fact due to relaxation of the EFG directions, which makes the problem even worse. Our hope is that understanding simultaneously Figs. 6, 7, and 8 will lead to a better understanding of G(r,t) for impurities in viscous liquids for t comparable to τ_n. The paper by J. H. Jensen[12] discussing similar measurements in a different liquid will clarify many of the ideas here cursorily presented.

CONCLUSIONS

The Mössbauer effect is an excellent tool for seeing phase charges, crystallization, melting, etc. and can make significant contributions to any such studies, where the appropriate nucleus can be used.

In at least the present glass-to-crystal transition, the short-range order hardly improves. Alternately, the nearest neighbors of the ferrous ion in aqueous glass are always H_2O, despite the absence of long-range order and the presence of the extra metastable energy.

Detailed study of the Mössbauer line shape may yet give additional information about atomic motions in liquids. It is already clear that the motions are too complicated for a simple diffusion model.

REFERENCES AND NOTES

1. H. Bernas and M. Langevin, J. Phys. 24, 1034 (1963).

2. A summary of the early work appeared in Hyperfine Structure and Nuclear Radiations, ed. by E. Matthias and D. Shirley, North-Holland Publishing Co., Amsterdam, 1968. A paper presented at this conference - I. Dezsi, Mössbauer Effect Studies on Frozen Solutions - contains much new materials and further references back to earlier work.

3. A. Nozik and M. Kaplan, J. Chem. Phys. 47, 2960 (1967). Another very useful paper by these authors is (Phys. Rev. 159, 2, 273 (1967) who point out here that $q_{lattice}$ and q_{ion} are not necessarily of opposite sign.

4. P. P. Craig and N. Stein, Phys. Rev. Lett. 11, 460 (1963).

5. D. C. Champeney and F. W. D. Woodhams, J. Phys. B. (Proc. Phys. Soc.) 1968, ser. 2, Vol. 1.

6. J. A. Elliott, H. E. Hall, and D. St. P. Bunbury, Proc. Phys. Soc. 89, 595 (1966).

7. An early reference is J. G. Mullen and R. C. Knauer, Mössbauer Effect Methodology, ed. Gruverman (Plenum Press, N. Y. 1969, Vol. 5). His student, A. A. Abras, in a recent Ph.D. thesis at Purdue makes a fine summary of the more recent work.

8. R. L. Cohen and K. W. West, Chem. Phys. Letters 13, 482 (1972).

9. W. A. Mundt and T. Sonnino, J. Chem. Phys. 50, 3127 (1969).

10. S. Bukshpan, C. Goldstein, and T. Sonnino, J. Chem. Phys. 49, 12 (1968).

11. A. Simopoulos, H. Kostikas, D. Petrides, and H. Wickman, Chem. Phys. Letters 7, 615 (1970).

12. J. H. Jensen, Phys. Kondens, Material 13, 273 (1971).

13. J. A. Tjon and M. Blume, Phys. Rev. 165, 2, 456 (1968).

14. D. Turnbull and M. H. Cohen, J. Chem. Phys. 34, 120 (1960).

15. Misfit, a suggested goodness-of-fit parameter, is a normalized modification of χ^2 such that (1) its value is independent of the duration of the run, (2) "normal" instrumental inadequacies lead to Misfit values of $< 0.1\%$ for ^{57}Fe experiments, (3) the error in Misfit clarifies whether a small χ^2 results from (a) good theory or (b) poor data.

16. D. E. Polk, J. Non-Cryst. Solids 5, 365 (1971).

HIGH-RESOLUTION MÖSSBAUER SPECTROSCOPY WITH TANTALUM-181

G. Kaindl and D. Salomon

Lawrence Berkeley Laboratory
University of California
Berkeley, California 94720

I. INTRODUCTION

An improvement of the resolution that may be obtained with the Mössbauer method is of considerable interest. The few Mössbauer resonances with lifetimes in the microsecond region have therefore attracted special attention. These are the 93 keV γ transition of ^{67}Zn ($T_{1/2}$ = 9.3 μs), the 6.2 keV γ transition of ^{181}Ta ($T_{1/2}$ = 6.8 μs), and the 13.6 keV γ transition of ^{73}Ge ($T_{1/2}$ = 4.6 μs). Only for the first two cases has the Mössbauer resonance been observed up to now [1,2].

The ultimate resolution which may eventually be reached in a Mössbauer measurement of hyperfine interactions is mainly determined by the natural linewidth of the Mössbauer γ rays and by the size of the pertinent nuclear parameters. For studies of isomer shifts, magnetic dipole, and electric quadrupole interactions, these are the change of the mean-square nuclear charge radius $\Delta \langle r^2 \rangle$, the magnetic dipole moment μ, and the electric quadrupole moment eQ for both nuclear states involved. Since all of them are very large for the 6.2 keV γ resonance of ^{181}Ta [3-5], and comparatively much larger than those known or expected for the γ resonances of ^{67}Zn and ^{73}Ge, the ^{181}Ta resonance may be considered as the top candidate for high-resolution Mössbauer spectroscopy applied to the study of hyperfine interactions.

This high sensitivity, on the other hand, makes the

6.2 keV γ resonance extremely vulnerable to lattice imperfections or impurities, resulting in relatively serious experimental difficulties. That may be the main reason why, until recently, this resonance has not been applied to the study of hyperfine interactions. Most of the work done concentrated on the problem of observing the resonance at all [1,6].

After the initial success of Sauer et al. [4,6], who measured the magnetic splitting of the 6.2 keV γ rays in a longitudinal external magnetic field of only 1.45 kOe, the subject was further studied by two groups, in Munich and in Berkeley. The present paper is based on experimental results obtained by the Berkeley group, but we would like to point out that part of our results, especially the systematics of isomer shifts in transition metals, the electric quadrupole splitting in rhenium metal, and the TaC spectrum, have also been obtained by the Munich group [7].

The paper is divided in three main parts, dealing with magnetic dipole interactions, electric quadrupole interactions, and isomer shift studies. A brief experimental section in the beginning is devoted to the special technique appropriate for the ^{181}Ta γ resonance.

II. EXPERIMENTAL TECHNIQUE

The extreme sensitivity of the 6.2 keV γ resonance to lattice imperfections, as well as the low γ ray energy, and the unusual large ratios of lineshifts to linewidths, require some refinements of the usual Mössbauer technique which will be mentioned briefly. Further details are reported elsewhere [17].

A. Source and Absorber Preparation

Two advantageous experimental features of the ^{181}Ta resonance are the almost 100% natural abundance of ^{181}Ta, and the easily preparable source activity of ^{181}W ($T_{1/2}$ = 140 d). For the present experiments it was produced with high specific activity by neutron irradiation of 93% enriched ^{180}W metal in a thermal neutron flux of $3\cdot10^{15}$ n/cm^{2}s for periods of at least 30 days. All the sources used were prepared by diffusing the ^{181}W activity into high-purity, and in most cases single-crystal transition metals in high-vacuum (~ 10^{-8} Torr) and at temperatures up to 2500°C. Prior to diffusion, the ^{181}W activity, dissolved in concentrated HF-HNO_3, was dropped onto the electropolished metal discs and reduced in hydrogen atmosphere (with the exception of the Pd source). Induction heating was used in all cases to heat the metal discs.

The Ta metal absorber used was prepared from a high-purity Ta foil (99.996% pure), rolled to a thickness of 4.1 mg/cm^2, and annealed for 10 h at 2000°C in ultra-high vacuum at ~ 10^{-9} Torr.

One of the difficulties in studying Ta compounds is the necessity of using thin and holefree absorbers due to the large photoabsorption cross-section of Ta for the 6.2 keV γ rays (σ_{ph} = 330 cm^2/g). These were obtained by sedimentating the finely powdered compounds in a benzene-polystyrene solution on 0.25 mil thick mylar foils. Naturally this technique is limited to rather stable compounds.

B. Mössbauer Spectrometer

A conventional sinusoidal velocity spectrometer [8] was used for the Doppler shift measurements. Both the moved source and the fixed absorber were mounted in an evacuated space in order to prevent acoustic disturbances. The 6.2 keV γ rays were detected with an Ar/CH_4-filled proportional counter.

In cases with a large isomer shift/linewidth ratio, the experiments were performed with small solid angles in order to prevent excessive geometrical broadening of the lines, coupled with a corresponding decrease in the observed resonance effect. In these cases as many as 2048 channels were used for recording the spectra.

III. RESULTS AND DISCUSSION

A. Magnetic Splitting

The magnetic splitting of the 6.2 keV γ rays in an external magnetic field was measured with both a velocity drive and a "magnetic drive" [4], using a source of $^{181}W(\underline{W})$

and a Ta metal absorber [11]. The velocity spectra are shown in Fig. 1. The single-line spectrum is quite asymmetric due to the interference effect between photoelectric absorption and nuclear resonance absorption, followed by internal conversion [9,10]. This effect was first observed by Sauer *et al.* [4]. For the magnetically split spectrum, the source was attached onto the flat surface of a cylindrically shaped Sm-Co permanent magnet, producing a longitudinal magnetic field of 2.93±0.03 kOe at the position of the source.

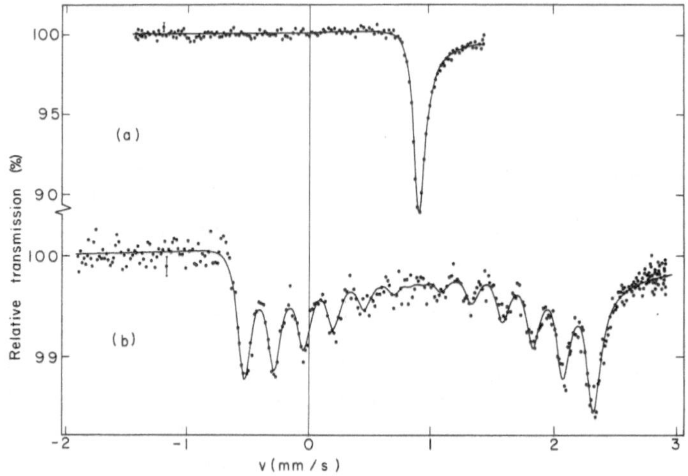

Fig. 1. Velocity spectra of the 6.2 keV γ rays, measured with a $^{181}W(\underline{W})$ source and a single line Ta metal absorber: (a) unsplit source, (b) source in a longitudinal magnetic field of 2.93±0.02 kOe.

The spectra were fitted with dispersion modified Lorentzian lines of the form

$$N(v) = N(\infty) - A(1 - 2\xi X)/(1 + X^2)$$

with $X = 2(v - v_0)/W$. Here $N(v)$ is the intensity transmitted at relative velocity v, v_0 is the position of the line, W is the full linewidth at half-maximum, and A is the amplitude of the line. The parameter ξ determines the relative magnitude of the dispersion term. According to Trammel *et al.* [9] the maximum value for ξ is given by

$$\xi = \left(\frac{\alpha\,\sigma_e'}{6\pi\,\lambda\!\!\!^{-2}}\right)^{1/2}$$

where σ_e' is the partial cross section for E1 photoelectric absorption, α is the internal conversion coefficient, and $\lambda\!\!\!^{-}$ is the wavelength of the incident γ rays, divided by 2π.

The results obtained from the analysis of the different spectra, including those of the magnetic drive spectrum of Ref. 11, are accumulated in Table 1. The best experimental halfwidth obtained (W/2 = 0.057±0.001 mm/s), uncorrected for finite absorber thickness, corresponds to about 17 times the natural linewidth. However, a maximum resonance effect of 11%, uncorrected for background, has been observed, indicating that a considerable amount of the broadening is due to the very thick Ta metal absorber (effective absorber thickness t = 23).

Table 1. Summary of experimental results obtained with a Ta metal absorber and an unsplit and magnetically split ^{181}W(W) source, respectively.

	Velocity spectra		Magnetic drive spectrum
	unsplit	magnetically split	
W/2 (mm/s)	0.056±0.001	0.066±.002	0.069±0.003
2ξ	-0.31±0.01	-0.31±0.01	-0.35±0.07
IS·(mm/s)	0.857±0.005	0.854±0.005	---
g(9/2)/g(7/2)	---	1.78±0.02	1.76±0.04

The experimental value for 2ξ may be compared with theory. Using α = 46±8 [12], and σ_e' = $9.8\cdot10^4$ b [13], we obtain 2ξ = -0.31, in very good agreement with our experimental value of 2ξ = -0.31±0.01.

The g-factor ratio is given by the weighted average of the two measurements

$$g(9/2)/g(7/2) = 1.77 \pm 0.02 \quad .$$

With $\mu(7/2) = +2.35 \pm 0.01$ n.m. [14] we obtain for the magnetic moment of the 6.2 keV state

$$\mu(9/2) = +5.35 \pm 0.09 \text{ n.m.}$$

This value agrees with our earlier result [11], but is slightly larger than that of Ref. 4.

Recently, we have also observed the magnetic splitting of the 6.2 keV γ rays in the magnetic hyperfine field of ^{181}Ta impurities in Ni metal at room temperature [15]. From the resulting hyperfine spectrum a value for the magnetic field of ^{181}Ta in nickel at room temperature could be deduced: $H_{hf} = -89.6 \pm 1$ kOe. This represents the first application of the 6.2 keV γ resonance to the study of magnetic hyperfine interactions.

B. Electric Quadrupole Interaction

A high resolution may also be expected for the study of electric quadrupole interactions (EQI) because of the large electric quadrupole moment of the groundstate of ^{181}Ta, $Q(7/2) = +3.9 \pm 0.4$ b [3], and the large Sternheimer antishielding factor γ_∞ for the Ta ion, $\gamma_\infty \sim -60$ [16]. Initially, we observed the electric quadrupole splitting of the 6.2 keV γ rays with sources of ^{181}W diffused into high-purity single-crystal rhenium metal. By observing the emission spectrum parallel and perpendicular to the [0001]-direction of the single-crystal ^{181}W(<u>Re</u>)-sources, the assignment of individual transitions to the observed lines could be done unambiguously. From the completely resolved spectra, the sign and magnitude of the electric field gradient (EFG) at the ^{181}Ta site in rhenium metal, the ratio of the quadrupole moments of the excited state to that of the groundstate, and the isomer shift of ^{181}Ta(<u>Re</u>) relative to Ta metal could be derived [5]. A similar study, using a source of ^{181}W diffused into rhenium metal foils with preferred orientation has been undertaken by Wortmann [7].

The velocity spectra obtained with our single-crystal sources are presented in Fig. 2. As the 6.2 keV transition has pure E1 character with the spin sequence 9/2 → 7/2, the hyperfine spectrum resulting from an EQI alone consists of 11 hyperfine components, of which 7 are Δm = ±1 transitions, 3 are Δm = 0 transitions, and 1 is a mixed transition.

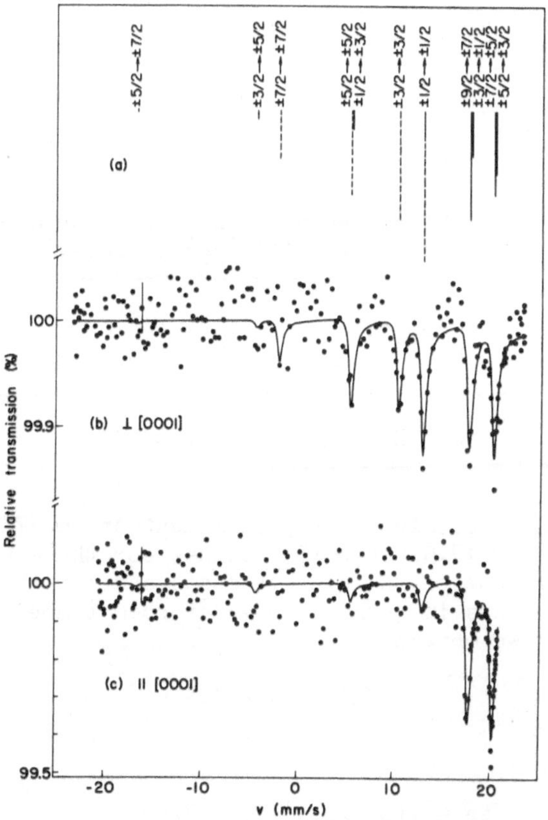

Fig. 2. Velocity spectra of the 6.2 keV γ rays of ^{181}Ta in rhenium metal versus a Ta metal absorber, with direction of observation (b) perpendicular and (c) parallel to the [0001]-axis. In (a) the positions and intensities of the individual components are given, in solid lines for the Δm = ±1 transitions, and in dashed lines for the Δm = 0 transitions as observed perpendicular to the [0001]-axis.

Both Mössbauer spectra were simultaneously fitted by least-squares with a superposition of dispersion-modified Lorentzian lines, where the amplitude of the dispersion term was set equal to $2\xi = -0.30$. As the only free parameters influencing the line positions, the z-component of the axially symmetric EFG eq, the ratio of the electric quadrupole moments Q(9/2)/Q(7/2), and the isomer shift IS were used.

The results of the least-squares fit analysis are presented in Table 2. For both spectra the total resonance effects, summed over all lines, were only ~ 1%, reflecting the considerable linebroadening observed. Only a small fraction of this (~ 0.1 mm/s for the solid angle $\Omega = 4\pi/500$ used) can be caused by geometrical broadening.

Table 2. Results of analysis of the hyperfine spectra of the single-crystal ^{181}W(<u>Re</u>) sources.

isomer shift (mm/s)	$\frac{Q(9/2)}{Q(7/2)}$	$e^2qQ(7/2)$ (10^{-6} eV)	Linewidth (mm/s)
-14.00±0.10	1.133±0.010	-2.15±0.02	0.60±0.04

With the electric quadrupole moment of the 7/2+ groundstate, Q(7/2) = +3.9±0.4 b [3], values for the electric quadrupole moment of the 6.2 keV state and for the EFG eq at the ^{181}Ta nucleus in rhenium metal at room temperature can be derived. We obtain

$$Q(9/2) = +4.4\pm0.5 \text{ b}$$

and

$$eq = -(5.5\pm0.5)\cdot10^{17} \text{ V/cm}^2 \quad .$$

The groundstate and the 6.2 keV state of ^{181}Ta have been classified as intrinsic proton states with the Nilsson assignments 7/2+[404] and 9/2-[514], respectively. Neglecting bandmixing, we obtain within the framework of the Nilsson model from our ratio of the spectroscopic quadrupole moments a value for the ratio of the intrinsic quadrupole moments $Q_0(9/2)/Q_0(7/2) = 0.969\pm0.009$, indicating that the deformation of the 9/2-[514] state may be slightly smaller than that of the 7/2+[404] groundstate.

Using the same experimental technique as for the rhenium experiment, we have additionally measured the hyperfine spectra of the 6.2 keV γ rays of ^{181}Ta impurities in hafnium, osmium and ruthenium metal at room temperature [18]. Some typical velocity spectra are shown in Fig. 3: in (a) for a polycrystalline ^{181}W(Os) source, and in (b) for a single-crystal ^{181}W(Ru) source, with direction of observation perpendicular to the [0001]-direction. The data were analyzed similarly as the rhenium spectra, but in these cases the ratio of the quadrupole moments Q(9/2)/Q(7/2) = 1.133±0.010 was additionally kept constant during the fit procedures.

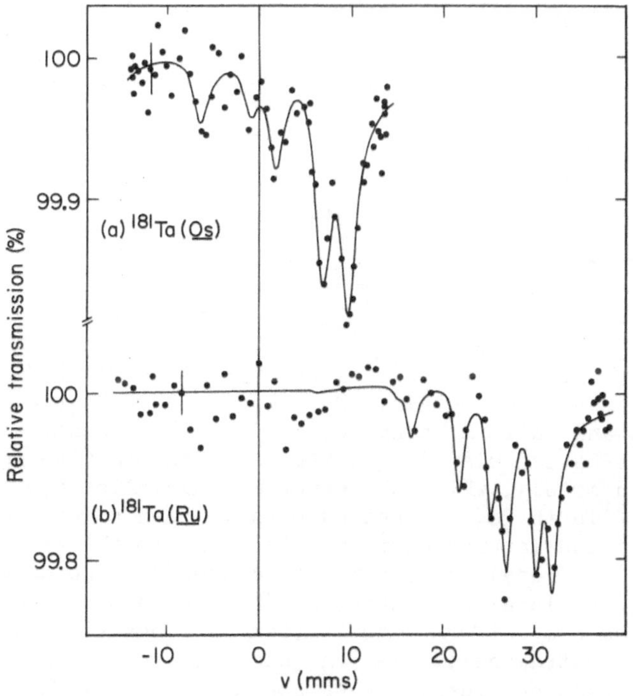

Fig. 3. Mössbauer absorption spectra for sources of ^{181}W(Os) (polycrystalline) (a) and ^{181}W(Ru) (single-crystal, observed perpendicular to the [0001]-axis) (b), both analyzed with a single-line absorber of Ta metal.

The experimental results for the EFGs, including those for rhenium metal, are summarized in Table 3. For all the studied cases the sign of eq was also determined. Our value for the EFG at ^{181}Ta in rhenium metal compares well with the results of nuclear specific heat [19] and nuclear acoustic resonance [20] measurements for pure rhenium metal. Using the quadrupole moment of the groundstate of ^{185}Re (Q(^{185}Re) = +2.3±0.9 b [21]), a value of eq = $-(4.9\pm1.9)\cdot10^{17}$ V/cm^2 can be derived for the EFG in rhenium metal at 4.2 K. This value is in good agreement with our room temperature result for ^{181}Ta(Re), though of less accuracy due to the large error of Q(^{185}Re).

Table 3. Electric quadrupole interaction of ^{181}Ta impurities in hexagonal transition metals at room temperature.

host	$e^2qQ(7/2)$ (10^{-6} eV)	eq (10^{17} V/cm^2)	W (mm/s)
Re	-2.15±0.02	-5.5±0.5	0.60±0.04
Os	-2.35±0.04	-6.0±0.7	1.8±0.2
Hf	+1.83±0.10	+4.7±0.7	1.6±0.4
Ru	-1.56±0.04	-4.0±0.5	1.3±0.2

Presently, no satisfactory theory exists for EFGs in hexagonal transition metals. Our results show that a simple "point-ion and uniform background model" [22] is not adequate, since the EFGs calculated with its help are positive for all the studied metals, while the experimental values are negative for ^{181}Ta in Re, Os, and Ru metal. Such a model considers only the lattice contribution q_{latt} to the EFG. Obviously, the local contributions, caused by a non-cubic arrangement of localized charge around the central atom as well as by a non-uniform distribution of conduction electrons within the central cell, play an essential role. The latter effect has been considered by Watson et al. [23], who predicted an "overshielding effect", caused by the conduction electrons, especially in cases with a high density of states at the Fermi energy. According to this theory, the local conduction electron contributions should be linearly related to the lattice field gradient, $q_{loc} = r\cdot q_{latt}$, so that we may write for the total EFG

$$q = q_{latt}[(1 - \gamma_\infty) + r(1 - R_Q)] \quad ,$$

where γ_∞ and R_Q are the lattice and atomic Sternheimer factors, respectively. With $\gamma_\infty \sim -60$ [16] and $R_Q \sim -0.2$ [24] for Ta, and using q_{latt} values as calculated with Ref. 22, we may derive overshielding factors r of -160, -85, and -75 for ^{181}Ta in Re, Os, and Ru, respectively. These values are not too different from those predicted for the pure metals, $r \sim -100$ for Re, and $r \sim -55$ for Os [23]. The EFG for ^{181}Ta(<u>Hf</u>), on the other hand, is positive, and even larger that $q_{latt}(1 - \gamma_\infty)$, and its temperature dependence, measured by TDPAC of the 5/2+ state at 482 keV of ^{181}Ta [25], is well described by the anisotropic variation of the lattice parameters with temperature. This means that in this case the lattice contribution may play an essential role.

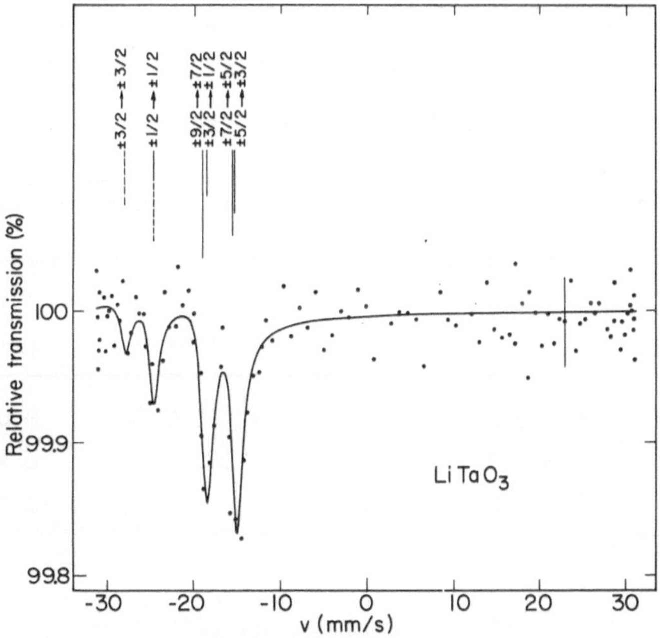

Fig. 4. Hyperfine splitting of the 6.2 keV γ rays in $LiTaO_3$ at room temperature. The position and intensities of the individual components are indicated by solid lines for the $\Delta m = \pm 1$ transitions and by dashed lines for the $\Delta m = 0$ transitions.

We have also applied the 6.2 keV γ resonance to the study of EQI in Ta compounds, namely $LiTaO_3$ and $NaTaO_3$ [26]. At room temperature $LiTaO_3$ is hexagonal, while $NaTaO_3$ is rhombic, with only small deviations from a tetragonal symmetry. The absorbers were prepared from the finely powdered compounds by a sedimentation technique. The resulting hyperfine spectrum for $LiTaO_3$, measured with a $^{181}W(\underline{W})$ source, is shown in Fig. 4. In both cases the absorber preparation technique was found to result in a highly preferred orientation of the axis of the EFG perpendicular to the absorber planes. Therefore, the spectrum shown in Fig. 4 was measured with the absorber plane tilted by 45° to the γ ray direction, in order to get an appreciable intensity in the ±1/2 → ±1/2 component, which is necessary for a determination of the sign of the EFG.

Both spectra were fitted under the assumption of an axially symmetric EFG, even though this assumption is exactly valid only for $LiTaO_3$. Both the amplitude of the dispersion term and the ratio of the electric quadrupole moments were kept constant during the fit procedures. The results are given in Table 4, where for completeness also the results for $KTaO_3$ are included.

Table 4. Summary of experimental results for $LiTaO_3$, $NaTaO_3$ and $KTaO_3$. The isomer shift is given relative to a Ta metal absorber.

Compound	IS (mm/s)	$e^2qQ(7/2)$ (10^{-6} eV)	eq (10^{17} V/cm^2)	W/2 (mm/s)
$LiTaO_3$	-24.0±0.3	+2.75±0.06	+7.05±0.80	0.8±0.1
$NaTaO_3$	-13.3±0.3	-1.02±0.07	-2.61±0.45	0.4±0.1
$KTaO_3$	-8.2±0.2	--	--	0.8±0.1

C. Isomer Shifts

The unique features of the 6.2 keV γ resonance among all other Mössbauer transitions are most clearly expressed by the isomer shift results. Until very recently, the only lattices in which the ^{181}Ta γ resonance had been observed, were W and Ta metal [6]. Then, at about the same time, Wortmann reported on an observation of the resonance for ^{181}Ta(Pt) [27], and the authors published their isomer results for ^{181}Ta impurities in Pt, Ir, Nb, and Mo, and $KTaO_3$ [28].

We have now observed isomer shifts for ^{181}Ta impurities in most of the transition metal hosts, and a few Ta compounds. The isomer shifts cover a total range of 110 mm/s, corresponding to 17,000 times twice the natural linewidth ($W_0 = 2\hbar/\tau = 6.436 \cdot 10^{-3}$ mm/s), or 1,600 times the minimum

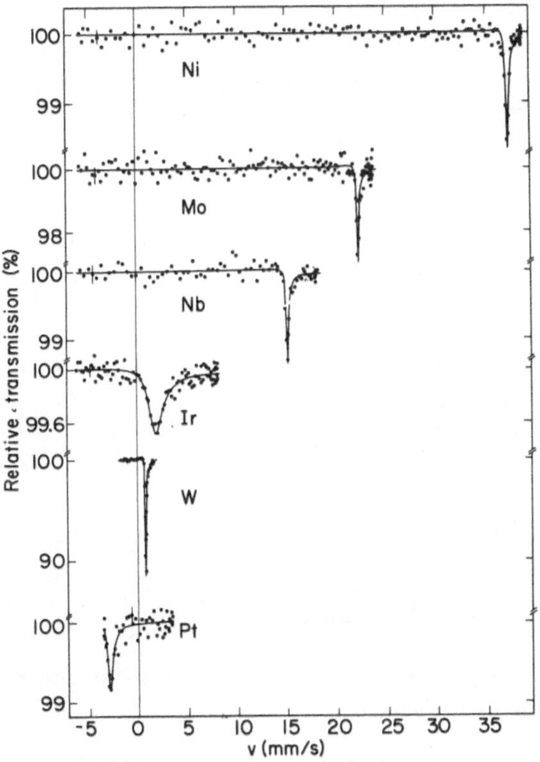

Fig. 5. Single-line spectra for sources of ^{181}W diffused into single-crystal transition metal hosts, analyzed with a Ta metal absorber. The spectrum for the Ni source was measured at 412°C, while all the other spectra were recorded at room temperature.

experimental linewidth obtained up to now [27]. These lineshift to linewidth ratios are more than an order of magnitude bigger than for any other Mössbauer transition presently in use, and the sensitivity of the lineposition of the 6.2 keV γ resonance to subtle solid state effects may therefore be considered as rather unique.

C.1. Systematics of isomer shifts. The isomer shift results reported here were derived from single-line and split spectra. Some representative single-line spectra for metallic sources are shown in Fig. 5. The large range of the isomer shifts is clearly exhibited.

We have also studied isomer shifts for a few compounds, namely $LiTaO_3$, $NaTaO_3$, $KTaO_3$ and TaC [29]. Some typical spectra, all of them recorded with a $^{181}W(\underline{W})$ source, are presented in Fig. 6. So far, the largest isomer shift relative to Ta metal was observed for TaC.

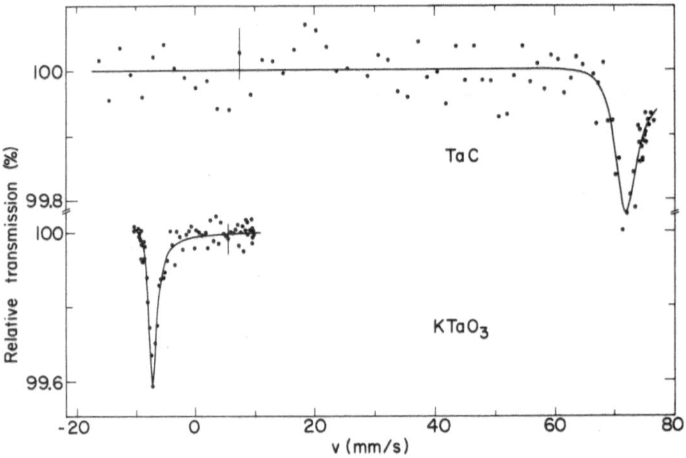

Fig. 6. Single-line absorption spectra for Ta compounds: (a) TaC, and (b) $KTaO_3$.

Table 5 gives a compilation of isomer shift results for both sources and absorbers at room temperature. The value quoted for the Ni lattice was extrapolated from the temperature dependence of the IS, measured for this host metal above the Curie temperature of nickel, taking into account the known temperature dependence of the lattice parameters of nickel.

Table 5. Compilation of isomer shift results, with source and absorber at room temperature. The isomer shifts are given relative to a Ta metal absorber.

Source Lattice	IS (mm/s)	W(FWHM) (mm/s)	Effect (%)
V	-33.6±0.7	7.7±2.6	0.1
Ni	-39.5±0.2	0.50±0.08	1.6
Nb	-15.26±0.10	0.19±0.06	1.5
Mo	-22.50±0.10	0.22±0.03	3.0
Ru	-27.50±0.30	1.3±0.2	0.7
Rh	-28.80±0.25	3.4±0.5	0.3
Pd	-27.55±0.25	2.7±0.6	0.3
Hf	-0.60±0.30	1.6±0.4	0.2
Ta	-0.075±0.004	0.184±0.006	2.4
W	-0.860±0.008	0.112±0.002	11.3
Re	-14.00±0.10	0.60±0.04	1.3
Os	-2.35±0.04	1.8±0.2	0.8
Ir	-1.84±0.04	1.60±0.14	0.5
Pt	+2.66±0.04	0.30±0.08	0.9
Absorbers			
$LiTaO_3$	-24.04±0.30	1.6±0.2	0.9
$NaTaO_3$	-13.26±0.30	1.0±0.2	0.9
$KTaO_3$	-8.11±0.15	1.5±0.2	0.3
TaC	+70.8±0.5	2.4±0.4	0.2

The observed experimental linewidths range from 17 times twice the natural linewidth for the W source up to 1200 times for the V source. The narrowest linewidths are observed for those host metals which are continuously miscible with Ta metal (with the exception of V). The best lineshift to linewidth ratios are observed for the Mo, Nb and Ni sources, even though the experimental linewidths for these three host metals are up to 77 times the natural one (in the case of nickel). Some of our results, namely the isomer shifts for the host metals V, Mo, Pd, W, Re and Pt, as well as the one for TaC have also been obtained by Wortmann [7,27], and his results are in good agreement with ours.

The isomer shifts for transition metal hosts exhibit some interesting systematic features if plotted versus the number of outer electrons for the series of 3d, 4d and 5d transition metals, as done in Fig. 7. As one can see, they arrange themselves in 3 different groups corresponding to the 3d, 4d and 5d host metals. Without exception, the transition energy decreases when going from a 5d to a 4d and

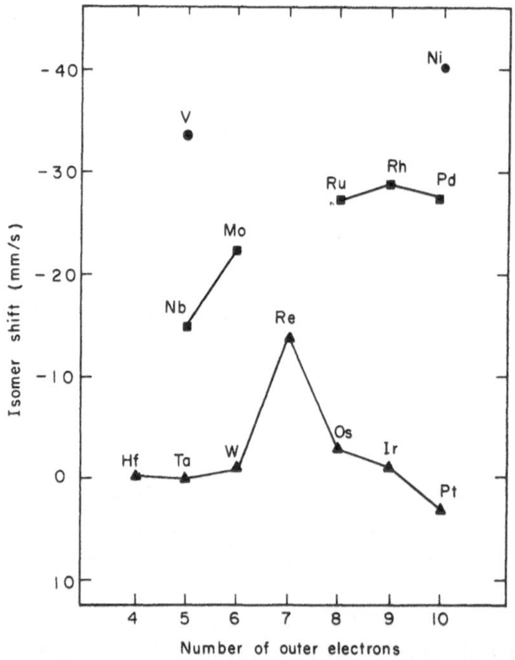

Fig. 7. Systematics of isomer shifts for transition metal hosts.

further to a 3d host element in a vertical column of the periodic system. The same behaviour has previously been observed for the ISs of the 14.4 keV γ rays of ^{57}Fe [30], and more recently by the authors for those of the 77 keV γ rays of ^{197}Au and the 90 keV γ rays of ^{99}Ru [15]. In these cases the changes of the mean square nuclear charge radii are relatively well established [32], at least their signs, so that we may derive from their systematics changes for the total electron density (L) within the nuclear volume. We find that for ^{57}Fe, ^{99}Ru, and ^{197}Au impurities in transition metals the electron density increases when proceeding from a 5d to a 4d and finally to a 3d host metal in a vertical column. The same conclusion can be drawn from less complete isomer shift data of the 73 keV γ rays of ^{193}Ir [33], the 99 keV γ rays of ^{195}Pt [34], and the 36 keV γ rays of ^{189}Os. If we assume the same dependence for the electron density at ^{181}Ta impurities we may conclude that $\Delta\langle r^2\rangle$ is negative for the 6.2 keV γ transition, since the 6.2 keV transition energy decreases when going from 5d to 4d, and finally to 3d host metals in a vertical column of the periodic system (Fig. 7).

From the isomer shifts measured for ^{193}Ir(<u>Ni</u>) and ^{193}Ir(<u>Pt</u>) [33], we derive an electron density difference of $\Delta L(Ir) = L(Ni) - L(Pt) = +3.5\cdot10^{26}$ cm^{-3}, using $\Delta\langle r^2\rangle = 4.6\cdot10^{-3}$ fm^2 [32]. Similarly, we get from our measurements for ^{197}Au(<u>Ni</u>) and ^{197}Au(<u>Pt</u>) [15], using $\Delta\langle r^2\rangle = 1.5\cdot10^{-2}$ fm^2 [35], a value of $\Delta L(\overline{Au}) = 2.7\cdot10^{26}$ cm^{-3}. If we take the average of the two values and correct for relativistic effects [36], we get $\Delta L(Ta) \sim 2.5\cdot10^{26}$ cm^{-3} between the nickel and platinum hosts. With an isomer shift difference of 42 mm/s between these hosts, we finally deduce

$$\Delta\langle r^2\rangle \sim -1.6\cdot10^{-2}\ \text{fm}^2$$

as an estimate for the change of the mean-square nuclear charge radius of the 6.2 keV transition.

This sizable $\Delta\langle r^2\rangle$, combined with the narrow line-width and the high atomic number of Ta forms the basis for the unique resolution obtainable in isomer shift studies with the 6.2 keV γ rays.

The above arguments, which have been used in a very similar way by Wortmann [7], should only be considered as an order-of-magnitude estimate for $\Delta\langle r^2\rangle$. The absolute value of $\Delta \langle r^2 \rangle$ may be subject to an appreciable error, but the sign is rather well established. The negative sign for $\Delta \langle r^2 \rangle$ is favoured by the ratio of the intrinsic quadrupole moments $Q_o(9/2)/Q_o(7/2) = 0.969 \pm 0.009$, derived from our experimental ratio of spectroscopic quadrupole moments. Assuming constant volume for the nucleus in both nuclear states, and using a deformation parameter $\beta = 0.275$, we find that the excited state is slightly less deformed than the groundstate by $\Delta\beta = -0.007 \pm 0.0023$; the relation $\Delta \langle r^2 \rangle = 3/2\pi \cdot R_o^2\ \beta\ \Delta\beta$ leads then to a collective contribution to $\Delta \langle r^2 \rangle$ of $-(4.5^{+1.1}_{-1.6}) \cdot 10^{-2}$ fm^2.

All our isomer shift results including those for the compounds, are represented graphically in Fig. 8. The bonding in transition metal carbides is not yet well enough understood in order to allow a derivation of $\Delta \langle r^2 \rangle$ from the

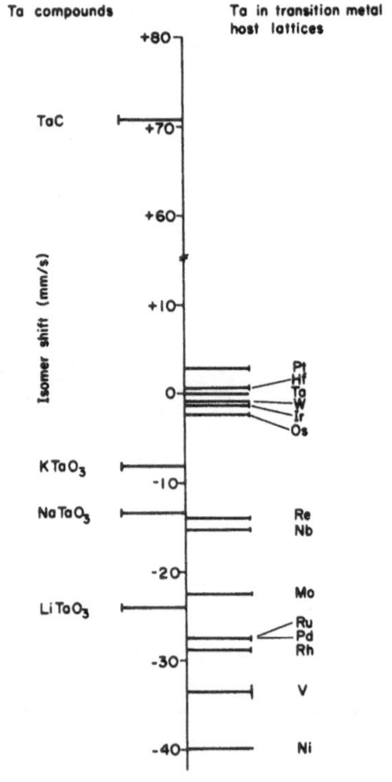

Fig. 8. Graphical representation of isomer shifts for the 6.2 keV γ rays of ^{181}Ta.

measured isomer shift of TaC. Especially, the question of the relative importance of metal-metal and metal-nonmetal bonding is not fully clarified [37-40].

C.2. Temperature dependence of the 6.2 keV gamma ray energy. Probably the best illustration for the high resolution of the 6.2 keV γ resonance is provided by the temperature dependence of linepositions for ^{181}Ta impurities in transition metals. The striking result is that the lineshifts are mainly caused by the temperature dependence of the isomer shift, and not as usually observed (for ^{57}Fe and ^{119}Sn) by the thermal redshift [41,42]. We find that for ^{181}Ta(Ni) the temperature shift of the line position is -33 times the thermal redshift expected for a Debye-solid at high temperature, and amounts to a shift of 2.3 natural widths per degree.

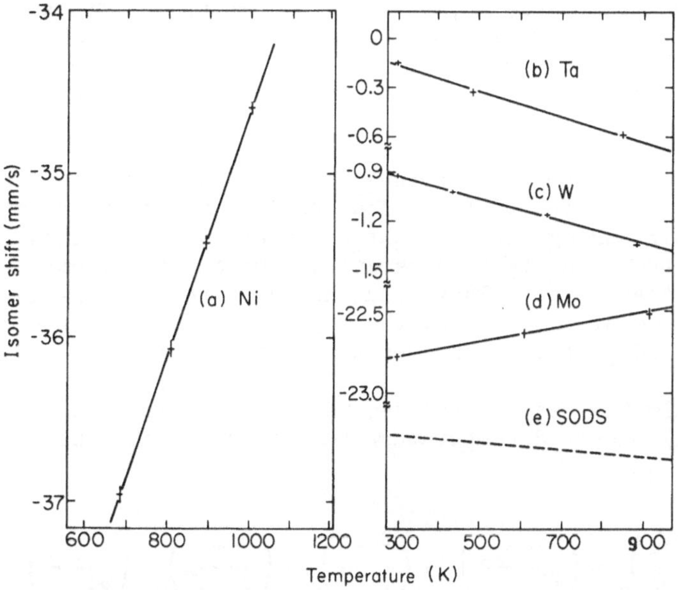

Fig. 9. Temperature dependence of line shifts (relative to a Ta metal absorber at room temperature) for sources of ^{181}W diffused into Ni (a), Ta (b), W (c), and Mo (d). For comparison, the second order Doppler shift expected for a Debye-solid at high temperature is also plotted in (e). All curves are plotted with the same scale on both axis.

The host metals investigated so far are Ta, W, Pt, Mo and Ni. For the experiments the moved sources were heated in a vacuum furnace, the temperature being measured by a Pt-Pt/Rh thermocouple directly in contact with the source metal discs. The fixed Ta metal absorber was kept at room temperature.

Some typical results are shown in Fig. 9. Within the accuracy of our measurements, the data can be described rather well by a linear relationship between lineshift S and temperature. The solid curves are the results of least-squares fits of straight lines to the data.

The influence of the thermal expansion of the lattice on the experimentally observed temperature dependence of the γ ray energy was first discussed by Pound et al. [43]. Combining the temperature shift results for iron metal [41] with the results of high pressure experiments on iron metal [43], they found that only ~ 8% of the temperature shift was caused by the thermal expansion of iron metal. The main part of the experimental shift could very well be described by the second order Doppler shift, assuming a Debye model for iron metal.

More recently, Rothberg et al. [44] have analyzed the temperature shifts of the resonance line for β-tin by additionally taking into account a temperature variation of the electron density at constant volume due to electron-phonon interaction. They find, however, that for β-tin, as in the case of iron metal, the thermal redshift is the dominating cause of the observed temperature shift.

Following Ref. 43 and 44, we may write for the experimentally observed temperature variation of the line-position S

$$\left(\frac{\partial S}{\partial T}\right)_P = \left(\frac{\partial S_{SODS}}{\partial T}\right)_P + \left(\frac{\partial S_{IS}}{\partial T}\right)_V + \left(\frac{\partial S_{IS}}{\partial \ln V}\right)_T \left(\frac{\partial \ln V}{\partial T}\right)_P$$

The first term represents the second order Doppler shift, which is given for a Debye-solid at high temperatures by $-3k/2Mc$ in cm/s, amounting to $-2.299\cdot 10^{-4}$ mm/s per degree for the 6.2 keV γ rays of ^{181}Ta. The second term represents the explicit temperature dependence of the isomer shift at constant volume due to electron density changes caused by the effects of electron-phonon interaction upon the electronic state of the conduction electrons. The third term describes the volume dependence of the isomer shift created by thermal expansion.

The experimental data are summarized in column 2 of Table 6. Our value for $\left(\frac{\partial S}{\partial T}\right)_P$ for the W host lattice agrees quite well with the result of Taylor et al. [45]. From the experimental lineshifts, values for the isobaric temperature dependence of the isomer shift $\left(\frac{\partial S_{IS}}{\partial T}\right)_P$ are derived after applying the correction for thermal redshift. The results are presented in column 3 of Table 6. While for nickel and molybdenum the γ ray energy increases with increasing temperature, the opposite effect is observed for Ta, W and Pt.

Table 6. Isobaric temperature variation of transition energy of the 6.2 keV γ rays, $\left(\frac{\partial S}{\partial T}\right)_P$, for ^{181}Ta impurities in transition metal hosts (column 2). In column 3 the derived fraction due to isomer shift, $\left(\frac{\partial S_{IS}}{\partial T}\right)_P$, is given, and in column 4 the isobaric thermal expansion coefficients extracted from x-ray data for the present temperature region (0-800°C).

host metal	$\left(\frac{\partial S}{\partial T}\right)_P$ (10^{-4} mm/s deg^{-1})	$\left(\frac{\partial S_{IS}}{\partial T}\right)_P$ (10^{-4} mm/s deg^{-1})	$\left(\frac{\partial \ln V}{\partial T}\right)_P$ (10^{-5} deg^{-1})
Ni	73.2±3.5	75.5±3.5	5.22
Mo	3.6±0.6	5.9±0.5	1.68
Ta	-8.0±0.5	-5.7±0.5	2.03
W	-7.1±0.2	-4.8±0.2	1.42
Pt	-17.6±0.9	-15.3±0.9	2.91

A separation of the isobaric temperature dependence of the isomer shift into an explicitly temperature dependent part and a volume dependent part can presently not be carried out. However, high-pressure isomer shift studies would directly yield the isothermal change with volume $\left(\frac{\partial S_{IS}}{\partial \ln V}\right)_T$, which then could be used to derive values for the explicit temperature coefficients. From the size of the isobaric temperature coefficients we may expect a rather high sensitivity of the transition energy of the 6.2 keV γ rays on hydostatic pressure.

In the case of ^{57}Fe, isomer shifts have been measured for the 14.4 keV γ rays as a function of pressure for ^{57}Fe impurities in a series of 3d, 4d, and 5d transition metal hosts [46]. The isothermal changes with volume were found to be positive in all cases, even though their absolute values showed large variations. The positive sign results from a decrease of the total electron density at the nucleus with increasing volume. Furthermore, the experimental values for $\left(\frac{\partial S_{IS}}{\partial \ln V}\right)_T$, when plotted versus the isomer shifts of the host metals, exhibit a systematic increase with decreasing isomer shift or increasing electron density L, which is not unexpected: a simple scaling of L with the volume, taking into account the opposite contributions of s- and d-like conduction electrons, would result in such a correlation.

A similar behaviour may be expected for the ^{181}Ta case. It would be in agreement with the negative sign for $\Delta \langle r^2 \rangle$ and the observed isobaric temperature shifts. For a qualitative discussion we use a simple scaling assumption [47]

$$\Delta L = - \frac{\Delta V}{V} L_{6s}$$

taking $L_{6s} \sim 2\cdot10^{27}$ cm^{-3} for the 6s-density at the Ta nucleus [48]. Then we expect from the thermal expansion of the nickel lattice a value of $\left(\frac{\partial S_{IS}}{\partial \ln V}\right)_T \left(\frac{\partial \ln V}{\partial T}\right)_P \sim 170\cdot10^{-4}$ mm/s deg^{-1}, using $\Delta \langle r^2 \rangle = 1.6\cdot10^{-2}$ fm^2. This is more than twice as large as the experimentally observed isobaric variation with temperature. But we have to consider that the simple scaling assumption is probably overestimating the change of electron density with volume [46,47], and that the explicit temperature dependence is expected to be negative for ^{181}Ta. With increasing temperature the electron phonon-interaction should cause a small d $\rightarrow$ s electron transfer, thus increasing the electron density linearly with temperature [31,44]. Since the observed isobaric temperature variations of the ISs are positive for Ni and Mo, and negative for Ta, W, and Pt, we may expect $\left(\frac{\partial S_{IS}}{\partial T}\right)_V$ to be of a comparable size as $\left(\frac{\partial S_{IS}}{\partial \ln V}\right)_T \left(\frac{\partial \ln V}{\partial T}\right)_P$. If we assume, for an order of magnitude estimate, a transfer of $\sim 10^{-5}$ d-electrons per degree [44], we get $\left(\frac{\partial S_{IS}}{\partial T}\right)_V \sim -21 \cdot 10^{-4}$ mm/s deg^{-1} (using $\Delta \langle r^2 \rangle = -1.6 \cdot 10^{-2}$ fm^2). This effect should depend very critically on the bandstructure.

With this picture the observed temperature variations of the IS can be understood qualitatively: the volume shift is expected to increase with decreasing isomer shift (from 5d to 3d), therefore resulting in positive overall temperature shifts for Ni and Mo. For the 5d elements, the explicit temperature variation is obviously dominant, resulting in small negative overall temperature shifts.

IV. CONCLUSION

The 6.2 keV γ resonance of ^{181}Ta can no longer be considered as an exotic case with limited applications. If experimental problems can be surmounted, its superiour resolution may be of great use for the study of subtle solid-state effects. This was demonstrated especially by the temperature shift results, where the effects caused by hyperfine interaction were found to be dominating over the thermal redshift. One should keep in mind that the present resolution is obtained with experimental linewidths typically of the order of 20 to 100 times the natural linewidth. Even though many applications of its unique features may be achieved with the present linewidths, an improvement is definitely of great interest.

The authors would like to thank Prof. D. A. Shirley for his constant interest in this work and valuable discussions.

REFERENCES

1. W. A. Steyert, R. D. Taylor, and E. K. Storms, Phys. Rev. Letters 14, 739 (1965).
2. H. de Waard and G. J. Perlow, Phys. Rev. Letters 24, 566 (1970).
3. V. S. Shirley, "Table of Nuclear Moments", in Hyperfine Interactions in Excited Nuclei, G. Goldring and R. Kalish, eds., Gordon and Breach, New York (1971) p. 1255.
4. C. Sauer, E. Matthias, and R. L. Mössbauer, Phys. Rev. Letters 21, 961 (1968).
5. G. Kaindl, D. Salomon, and G. Wortmann, Phys. Rev. Letters 28, 952 (1972).
6. C. Sauer, Z. Physik 222, 439 (1969), and references therein.
7. G. Wortmann, Thesis, Technical University Munich (1971).
8. G. Kaindl, Thesis, Technical University Munich (1969).
9. G. T. Trammel and J. P. Hannon, Phys. Rev. 180, 337 (1969).
10. Yu. M. Kagan, A. M. Afanasév, and V. K. Vojtovetskii, JETP Letters 9, 91 (1969).
11. G. Kaindl and D. Salomon, Phys. Letters 32B, 364 (1970).
12. A. H. Muir, Jr., Nucl. Phys. 68, 305 (1965).
13. "Table of Mass Absorption Coefficients", Norelco Report, May-June (1962).
14. L. H. Bennett and J. I. Budnick, Phys. Rev. 120, 1812 (1960).
15. G. Kaindl and D. Salomon, this conference.
16. F. D. Feiock and W. R. Johnson, Phys. Rev. 187, 39 (1969).
17. D. Salomon, Thesis, University of California, Berkeley (1972).
18. G. Kaindl and D. Salomon, Phys. Letters 40A, 179 (1972).
19. P. E. Gregers-Hansen, M. Krusius, and G. R. Pickett, Phys. Rev. Letters 27, 38 (1971), and references therein.
20. J. Buttet and P. K. Baily, Phys. Rev. Letters 24, 1220 (1970).
21. J. Kuhl, A. Steudel, and H. Walter, Z. Physik 196, 365 (1966).
22. F. W. DeWette, Phys. Rev. 123, 103 (1961).
23. R. E. Watson, A. C. Gossard, and Y. Yafet, Phys. Rev. 140, A375 (1965).

24. R. Sternheimer, private communication (1972).
25. R. M. Lieder, N. Butler, K. Killig, K. Beck, and E. Bodenstedt, in Hyperfine Interactions in Excited Nuclei, G. Goldring and R. Kalish, eds., Gordon and Breach, New York (1971) p. 449.
26. D. Salomon and G. Kaindl, Bull. Am. Phys. Soc. 17, 681 (1972).
27. W. Wortmann, Phys. Letters 35A, 391 (1971).
28. D. Salomon, G. Kaindl, and D. A. Shirley, Phys. Letters 36A, 457 (1971).
29. The authors are indebted to Dr. N. H. Krikorian, Los Alamos, for the high-purity TaC.
30. S. M. Quaim, Proc. Phys. Soc. (London) 90, 1065 (1967).
31. T. Muto, S. Kobayasi, M. Watanabe, and H. Kozima, J. Phys. Chem. Solids 23, 1303 (1962).
32. G. K. Shenoy and G. M. Kalvius, in Hyperfine Interactions in Excited Nuclei, G. Goldring and R. Kalish, eds., Gordon and Breach, New York (1971) p. 1201.
33. R. L. Mössbauer, M. Lengsfeld, W. von Lieres, W. Potzel, P. Teschner, F. E. Wagner, and G. Kaindl, Z. Naturforschung 26a, 343 (1971); F. E. Wagner, private communication.
34. D. Walcher, Z. Physik 246, 123 (1971).
35. M. Faltens and D. A. Shirley, J. Chem. Phys. 53, 4249 (1970).
36. D. A. Shirley, Rev. Mod. Phys. 36, 339 (1964).
37. L. E. Toth, Transition Metal Carbides and Nitrides, Academic Press, New York (1971).
38. E. Dempsey, Phil. Mag. 8, 285 (1963).
39. R. G. Lye and E. M. Logothetis, Phys. Rev. 147, 622 (1966).
40. V. Ern and A. C. Switendick, Phys. Rev. 137, A1927 (1965).
41. R. V. Pound and G. A. Rebka, Jr., Phys. Rev. Letters 4, 274 (1960).
42. B. D. Josephson, Phys. Rev. Letters 4, 341 (1960).
43. R. V. Pound, G. B. Benedek, and R. Drever, Phys. Rev. Letters 7, 405 (1961).
44. G. M. Rothberg, S. Guimard, and N. Benczer-Koller, Phys. Rev. 1B, 136 (1970).
45. R. D. Taylor and E. K. Storms, Bull. Am. Phys. Soc. 14, 836 (1969).
46. R. Ingalls, H. G. Drickhamer, and G. de Pasquali, Phys. Rev. 155, 165 (1967).
47. L. D. Roberts, D. O. Patterson, J. O. Thomson, and R. P. Levey, Phys. Rev. 179, 656 (1969).
48. Extrapolated from relativistic Dirac-Fock calculations for Os ions by J. B. Mann, Los Alamos, private communication.

EXPERIMENTS WITH THE MÖSSBAUER EFFECT IN ^{67}Zn †

Gilbert J. Perlow

Argonne National Laboratory

Argonne, Illinois 60439

A summary of the relevant properties of ^{67}Zn is given in Fig. 1. The problems are associated mainly with the narrowness of the line. The noise vibration velocities in most spectrometers are perhaps two orders of magnitude greater than the natural line width in zinc. Nevertheless it has been possible to obtain good spectra by successive improvements of the apparatus and the method.

The first work with ^{67}Zn in our laboratory at Argonne was done in collaboration with L. E. Campbell. We knew that the effect had been observed at Los Alamos by magnetic scanning (1) and that a Doppler spectrum had been obtained by Aksenov and collaborators(2) at the Lebedev Institute. It was clear from this that ZnO was a good first choice for source and absorber, and that the fine powder would need sintering to avoid macroscopic recoil. We assumed that any isomer shift due to chemical state would probably be fatal as the natural line width is only $2\Gamma_0 = 0.31$ microns/sec (or roughly one inch per day), and that would limit the scanning range to a few hundred times that amount -- still considerably less, for example, than the ^{57}Fe line width. The problem reduced to one of source and absorber preparation, and the modification or construction of a spectrometer. We had ^{66}ZnO available for source preparation by $^{66}Zn(d,p)^{67}Ga$ whose electron-capture decay leads to the 93-keV level in ^{67}Zn. We did not have enriched ^{67}Zn, but started with the natural isotopic abundance of 4.1%.

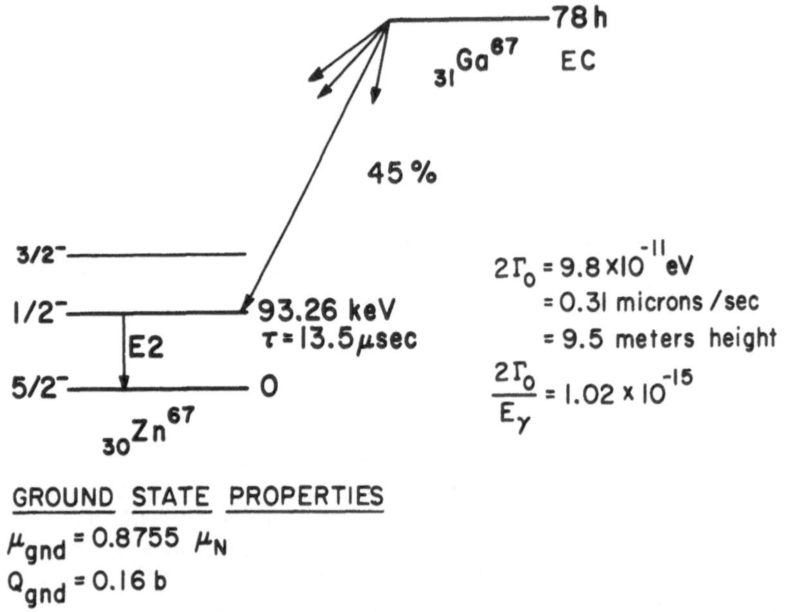

Fig. 1. Energy-level diagram.

Our first absorbers were sintered disks of about 2 electronic e-foldings in thickness. We were ignorant about how to anneal and only learned the technique when Hendrik De Waard and then Lawrence Conroy, a solid state chemist from Minnesota, joined me somewhat later. The first drive was one of our mechanical spectrometers, which employs a simple crank. We ran it with a very small crank throw and low angular velocity. The source and absorber were immersed in (noisily bubbling) liquid helium. We saw nothing. We then constructed an instrument in which the driving velocity was reduced with a divider consisting of two springs in series. The junction moves with velocity $v = v_0 \cdot k_1/(k_1 + k_2)$, where the k's are the respective spring constants. Again nothing was seen, probably because the lower spring (k_2) was not stiff enough. We replaced it with a piece of thin-walled aluminum pipe and saw our first, very bad, spectrum (Fig. 2). The "fit"

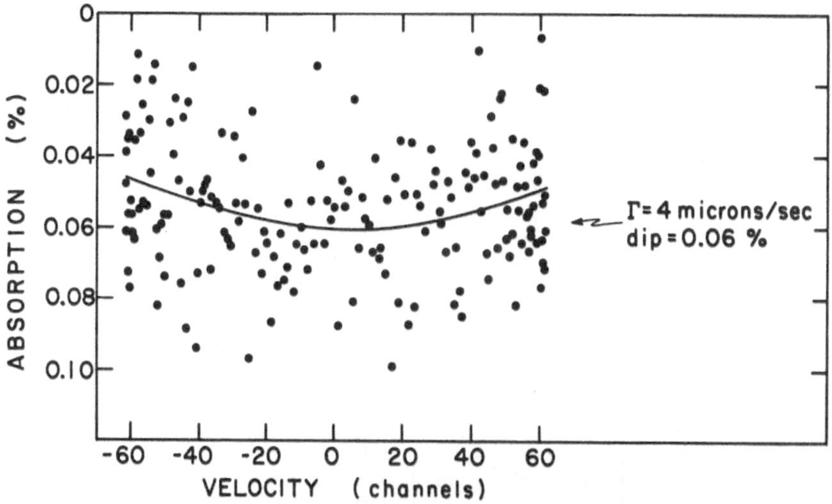

Fig. 2. Results with the stretched-pipe spectrometer.

shows a width of ~3.5 μm/sec, and a dip of 0.05%. The geometric effect, which is all this spectrum appears to illustrate, was in fact entirely negligible.

When Prof. DeWaard visited in the summer of 1969, a quartz-crystal piezoelectric drive was constructed. This is shown in Fig. 3. It contained ten x-cut quartz disks, each with a central hole. They were stacked interleaved with gold electrodes which were connected in such a way that the individual deflections added. The unit was placed in a sealed container kept at a reduced pressure of helium gas to reduce acoustic coupling to the bath. It was suspended by strings. The radiation emerged through a beryllium window, passed through the liquid helium and thence out through two more beryllium windows in the bottom of the dewar, and finally was detected in a Ge(Li) counter. The 93-keV line is quite clean; the background to be subtracted is initially about 30%, including a 7% correction for an unresolved γ ray. The background correction, which results in part from the decay of a 37-h contaminant of ^{69}Ge, becomes smaller as the source ages.

The drive was operated without feedback in a time-mode system. A sinusoidal voltage was placed across the

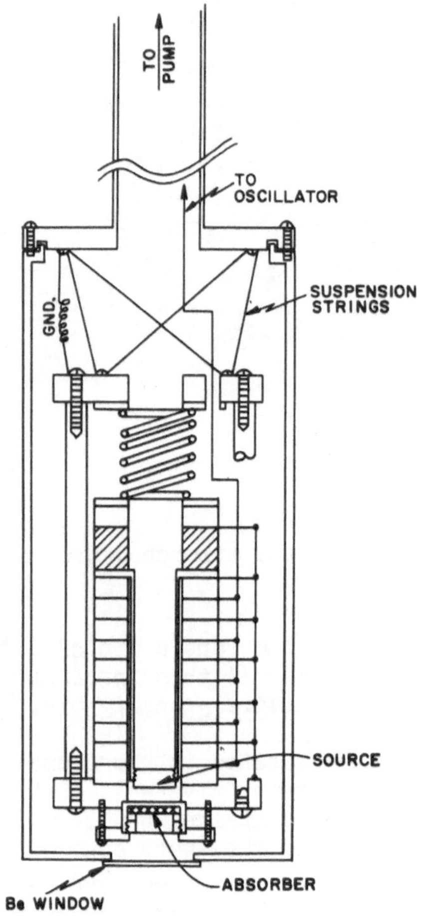

Fig. 3. Schematic diagram of the stacked-quartz spectrometer.

stack of crystals. Its zero crossings generated reset pulses and a train of address-advance pulses. The voltage was typically a few hundred volts at a few hundred hertz, well below the crystal resonance frequencies. The calibration depends primarily on the published constants for quartz. Typically the amplitude of source motion was about 50 Å.

The first respectable data taken with the drive are shown in Fig. 4. When the drive voltage was increased, we saw well-resolved hyperfine structure as in Fig. 5. There are in fact lines further out. They have been seen at Dubna(3), and we have also seen them. The absorber for our experiments was a sintered disk of ZnO enriched to 90% in ^{67}Zn and containing 1.57 g/cm^2 of ^{67}Zn.

The explanation of the structure is given in ref. (4), which I call DP. There is a quadrupole splitting of the $\frac{5}{2}^-$ ground state with separations ±5/2— (2E)— ±3/2—(E)—±1/2. The $\frac{1}{2}^-$ excited state is of course unsplit. Source and absorber have equal couplings. Because of the unoriented nature of source and absorber, there is isotropy in both emission and absorption.

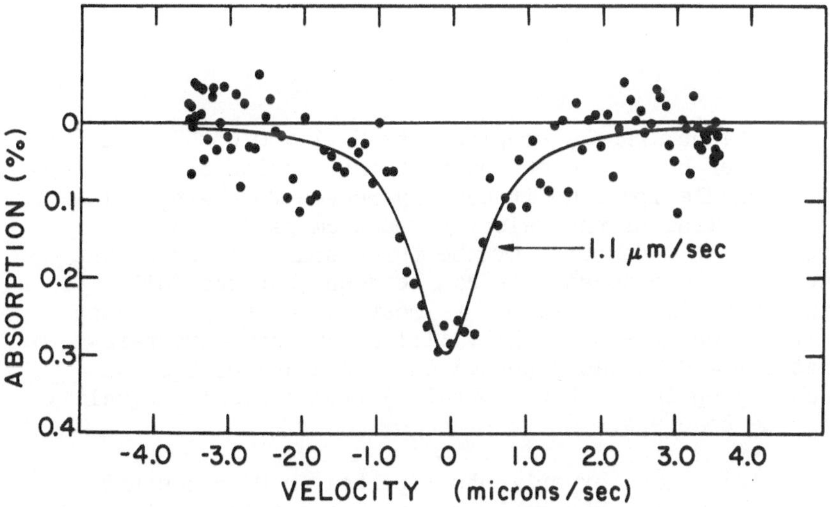

Fig. 4. Results with the apparatus of Fig. 3.

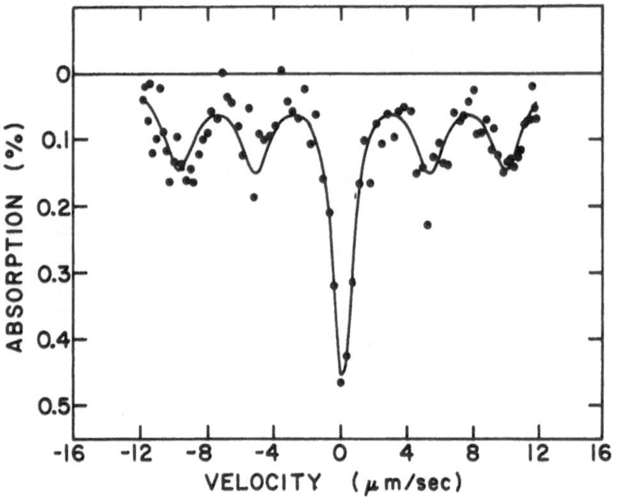

Fig. 5. Hyperfine structure in the ^{67}Zn spectrum.

The quadrupole coupling was 2.47 ± 0.03 MHz and was in rather better agreement with the results of a lattice-sum calculation than it had any reason for being. The central line of the velocity spectrum is due to the concurrent absorption of the three source lines by the absorber resonances. It can be seen that the full spectrum will consist of (almost) equally spaced dips with intensities (1:1:1:3:1:1:1). An asymmetry parameter $\eta = 0.23$ was reported in DP and $\eta = 0.19$ in the Dubna preprint. It was obtained from a small inequality in the spacings.

Additional experiments reported in DP concerned a small, but probably real, pressure-sensitive shift, and a search for the existence of a shift between ^{67}Zn in a ^{66}ZnO lattice and in a ^{67}ZnO lattice. No significant shift was seen -- a result in accord with prediction[5].

My main concern in the rest of the report is the measurement of the magnetic moment of the $\frac{1}{2}^-$ state by means of the nuclear Zeeman effect. The idea goes back to DeWaard's visit, but the bulk of the work was done in the last year in collaboration with L. Campbell and L. Conroy, with a later assist from W. Potzel.

I think it is clear to all that the use of single crystalline material greatly simplifies all magnetic spectra, and our first step was therefore to investigate this point. Enriched single crystalline ^{66}ZnO or ^{67}ZnO was considered too difficult to obtain. We settled for ordinary ZnO single crystals 1 cm in diameter and 0.5 mm thick, cut with c axis perpendicular to the faces. Because the Argonne cyclotron operates with α particles a large part of the time, it was convenient to make our sources indirectly by ^{64}Zn$(\alpha,n)^{67}$Ge $\rightarrow$ ^{67}Ga(78 h) or directly by ^{64}Zn$(\alpha,p)^{67}$Ga. Since ^{67}Ge decays with a half-life of 18 min, and the source preparation consumes a considerably longer time, the indirect process is equally as effective as the direct. The natural abundance of ^{64}Zn is 49%. The typical bombardment was 20 μA-h of 25-MeV alphas. The sources that resulted gave initial count rates in the window of perhaps 5000/sec. We found that a successful ritual for annealing them was to heat them in an oxygen atmosphere at 700°C for 12 hours and slowly cool them over the course of the next six. After bombardment, the sources were highly discolored but they became pale yellow upon annealing, just faintly more colored than before irradiation. A spectrum obtained with such a source is shown in Fig. 6. Since the γ ray is being observed along the direction of the hexagonal c axis, the only non-zero intensities correspond to transitions for which $\Delta m = \pm 1$. The intensity vanishes for $\pm\frac{1}{2} \rightarrow \pm\frac{5}{2}$ in the source. The seven lines then have intensities for E2 emission in the ratio of (1:2:1:3:2:0:0). The two outermost line positions are not within the range scanned. The agreement is found to be very good.

Since the pattern is not symmetric, the ordering of the lines immediately gives the sign of the quadrupole coupling. The coupling is positive. The quadrupole moment is also known (6) to be positive (+0.16b). Thus the field gradient is also positive. This agrees with the sign found in DP by a lattice sum. We have obtained

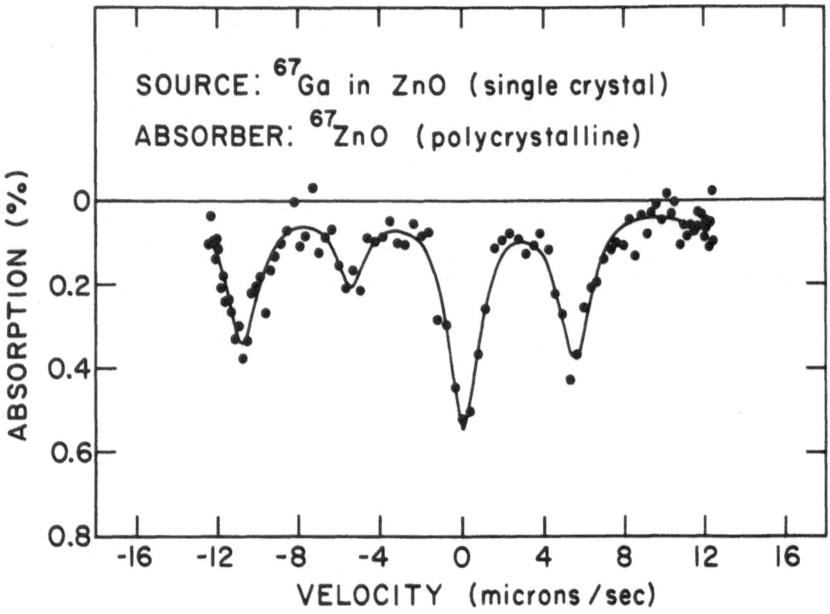

Fig. 6. Spectrum with single crystal source.

an asymmetry parameter from this and from a comparison run. It is $\eta = 0.09 \pm 0.04$. The quadrupole coupling is somewhat higher than in DP, $e^2qQ = 2.75$ MHz compared to 2.47. However, it is not necessarily a better value as there are instrumental problems.

From the area of the central line after background correction we can derive a value of Θ_M if we take equal characteristic temperatures for source and absorber. In addition, the result depends somewhat on how much of the observed width we wish to assign to the absorber. If $\Gamma_a = \Gamma_o$, we get $\Theta_M = 368°K$, while $\Gamma_a = \Gamma_{obs} - \Gamma_o$ gives $\Theta_M = 308°K$, and $\Gamma_a = \frac{1}{2}\Gamma_{obs}$ gives 315°K. These values are to be compared to Θ_D for the zinc atom alone, $\Theta_D \approx 290°K$ calculated in DP from crystallographic data(7), or $\Theta_D = 325°K$ from acoustic data (8). The point of all this is that we are observing substantially the full characteristic temperature in our experiments and we must therefore assume that the largest portion of our decays takes place from zinc atoms of normal (+2) valence in normal sites. It must be noted that even if there is a great deal of local trauma following the Auger cascades in the gallium decay, there is some modest fraction of 13.6 μsec available for repairs.

There may be a time-dependent variation of f_s with a characteristic time of perhaps a day. If so, it is a very interesting phenomenon, because with our relatively weak sources it might represent a form of self-damage that concentrates on undecayed ^{67}Ga sites. It is not too difficult to imagine such mechanisms, but the matter is highly speculative at this time.

The measurements of magnetic moment were done in external fields of 131 - 545 Oe, obtained from a small Alnico magnet. The field was varied by remagnetizing the Alnico. The construction is shown as an insert in Fig. 7. It screwed into the drive. The graph shows the variation of the field across the source area. This was obtained by traversing the latter with a Hall probe. The adequacy of the spatial resolution of the probe was determined by a traverse through a narrow air gap in a laboratory magnet. It can be seen that the single soft-iron pole face and the thin, magnetically soft, alloy disks produced a quite uniform field. H was measured by using the Hall probe as a transfer device. A correction of +3% was made for the effect of helium temperatures. The absorber was contained in a magnetic shield.

The determination of the magnetic moment is almost a null measurement because of a fortunate coincidence. This circumstance has allowed us to get a relatively good answer from not especially good data. It can be explained by reference to Fig. 8(a). If η is zero, the ground-state quadrupole Hamiltonian is diagonal in the natural quantization system; and if the field is applied along the symmetry axis, the Hamiltonian remains diagonal. The degenerate doublets with energy $E_Q\,(I_z^2)$ are split by the addition of $E_m = \pm|I_z|g_{gnd}\mu_N H$. That is, each acts like a separate spin $\frac{1}{2}$ nucleus with g factor proportional to $|I_z|$. If η is not negligible it is of course necessary to consider

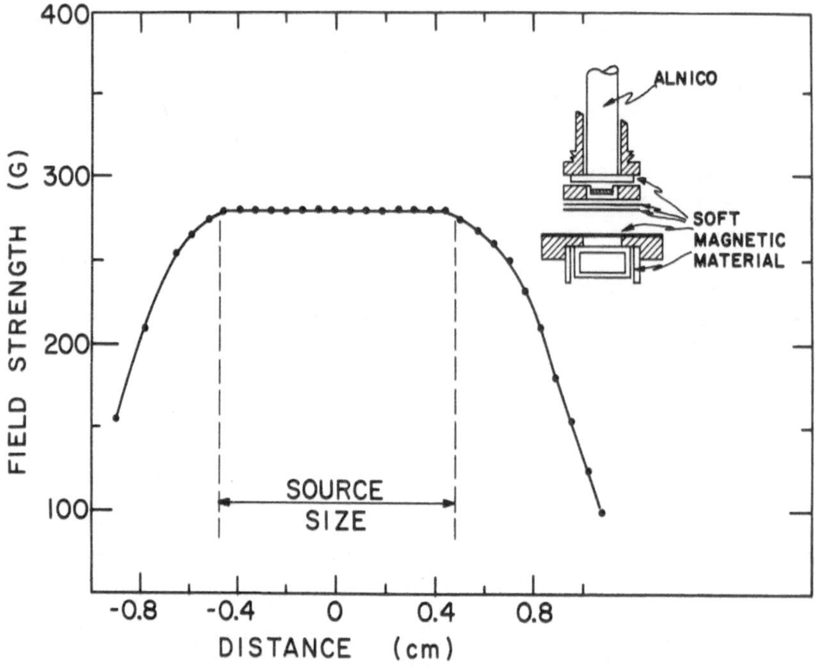

Fig. 7. Magnetic field variation across the source. The inset shows the construction of the source holder for the magnetic-moment measurements.

the entire matrix. The appropriate criterion is related to the size of η compared to the ratio of magnetic to quadrupole energies. We have drawn the diagram for the case of (expected) positive excited-state moment and (known) positive ground-state moment. Note that as the field increases, there is an increasing separation of the energies of the two transitions $(+\frac{1}{2} \rightarrow -\frac{1}{2})$ and $(-\frac{1}{2} \rightarrow +\frac{1}{2})$, here written in the form $(m_{exc} \rightarrow m_{gnd})$. On the other hand, the transitions $(+\frac{1}{2} \rightarrow +\frac{3}{2})$ and $(-\frac{1}{2} \rightarrow -\frac{3}{2})$ differ little in energy. In fact, the difference would be zero and independent of field if $g_{exc} = 3g_{gnd}$, i.e., if

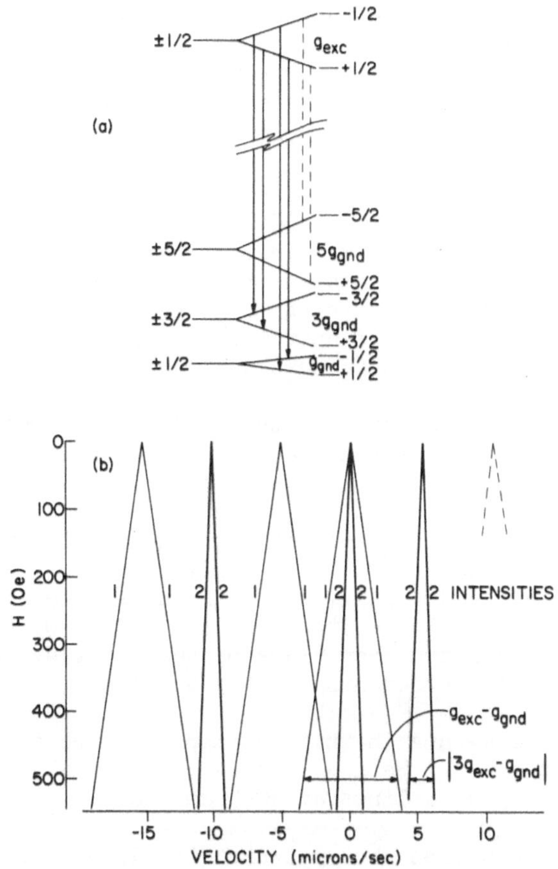

Fig. 8. Effect of combined quadrupole and magnetic interactions on the states involved in the 93-keV transition in ^{67}Zn. (a) The diagram is drawn for positive quadrupole coupling, positive magnetic moments in the ground and excited states, and a value of $\mu_{exc} = 0.6$ μ_N. (b) The velocity spectrum that results from the diagram at (a).

$\mu_{exc} = (3/5)\mu_{gnd}$ (including equality of sign). This turns out to be nearly the case. We observe the difference between $(+\frac{1}{2} \rightarrow +\frac{3}{2})$ and $(-\frac{1}{2} \rightarrow \frac{3}{2})$ as a slight broadening of a line. The broadening is then used as a correction to the assumption of exact equality. The sign of the correction is obtained by examining the other transition, whose small intensities and the resulting meager statistics do not lend themselves to a more active role in the measurement.

The spectrum that results is diagrammed in Fig. 8(b). Each line of the pair resulting from the transitions to $m_g = \pm\frac{3}{2}$ has unit intensity. This "$\frac{3}{2}$ pair" is close-spaced and is repeated three times in the full spectrum. The corresponding "$\frac{1}{2}$ pair" arising from transitions to $m_g = \pm\frac{1}{2}$ contains lines of half-unit intensity. They separate rapidly as the field is increased.

Fig. 9 shows a spectrum with 131 Oe on the source. We see the resolved $\frac{1}{2}$ pairs and a slight broadening of the base of the central 3/2 pair that corresponds to an unresolved 1/2 pair. The 3/2 pair at ~5 μm/sec, like the one at v = 0, is not noticeably broadened. The situation is more pronounced at 207 Oe, (Fig. 10), where one observes two resolved $\frac{1}{2}$ pairs and still no appreciable evidence for splitting of the $\frac{3}{2}$ pair. At 545 Oe (ig. 11), only the computer can keep track of the $\frac{1}{2}$ pairs, but we can start to see some broadening of the $\frac{3}{2}$ pairs.

We have five runs in which the spectra are of acceptable quality. We have observed the predicted crossover of $\frac{1}{2}$ and $\frac{3}{2}$ lines at ~2.5 μm/sec.

From perturbation theory, which is adequate for $\eta = 0.09$ and somewhat less so for $\eta = 0.23$, one finds that the effect of $\eta > 0$ is to decrease the splitting of the $\pm\frac{3}{2}$ doublet by the amount $2\eta^2A^2B/(A^2 - B^2)$, where $A = (3/40)e^2qQ$ and $B = (2/5)\mu_{gnd}H$. At 545 Oe this corresponds to 0.058 μm/sec if $\eta = 0.09$ and to 0.42 μm/sec if $\eta = 0.23$. If we accept and correct for the smaller value of η, the splitting in the two higher field runs (339 and 545 Oe) gives $|\mu_{exc} - (3/5)\mu_{gnd}| = 0.057\ \mu_N$. In all five runs, the 1/2 pair separation is consistent with $\mu_{exc} > (3/5)\mu_{gnd}$. We can therefore remove the absolute value signs, and from the known ground-state moment $\mu_{gnd} = +0.876\ \mu_N$ we get

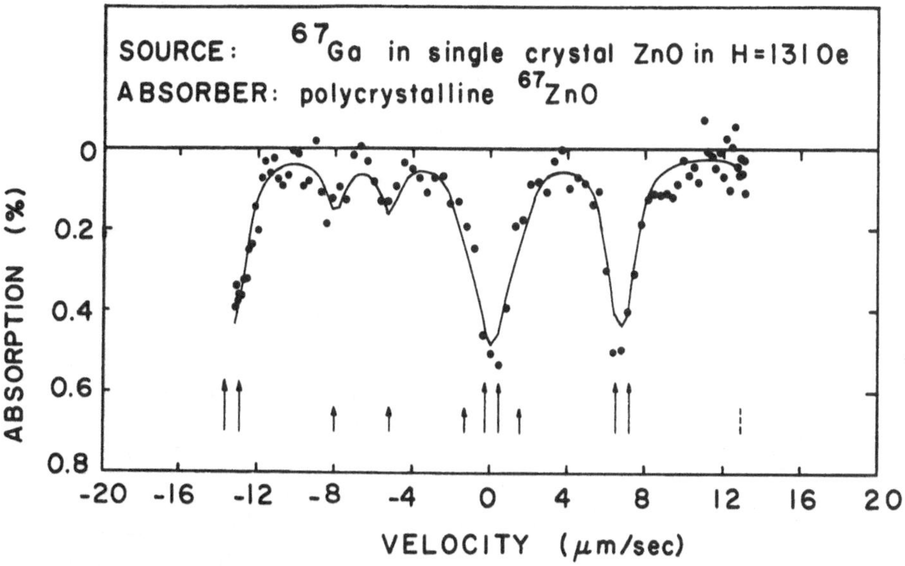

Fig. 9. Spectrum in 131 Oe. The arrows show the positions of the weak and strong lines.

$\mu_{exc} = (0.58 \pm 0.03)\mu_N$. We have assigned an error equal to half the asymmetry correction to account for our ignorance of this quantity.

There is evidence that ^{67}Zn is approximately described by the pairing model with Fermi energy lying between the single-neutron $f_{5/2}$ and $p_{1/2}$ levels. Some of this is found in the trend of spectroscopic factors measured for the Zn isotopes[9]. Another important point is that this E2 transition is retarded by a factor of 10 below single-particle (Moszkowski) estimates. E2 transitions are generally faster than such estimates because they are enhanced by static quadrupole deformations. However, in the pairing model, electric transition rates in odd-A nuclei are reduced by the factor $\{U_iU_f - V_iV_f\}^2$, where U_i^2 and V_i^2 are the fractional numbers of holes and

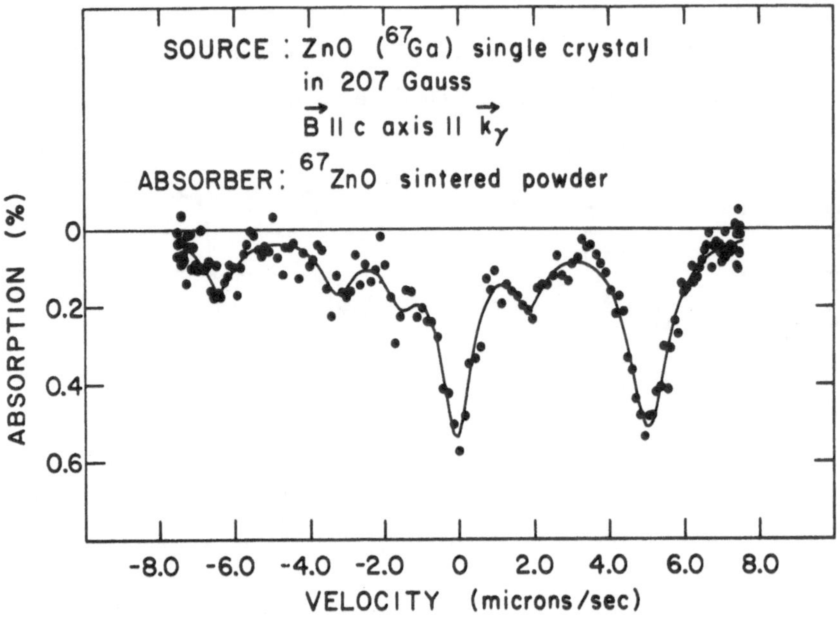

Fig. 10. Spectrum in 207 Oe.

particles, respectively, in the state i, and i → f in the transition[10]. If the Fermi energy lies halfway between the single-particle states, the bracket vanishes.

On this basis, the $\frac{1}{2}^-$ and $\frac{5}{2}^-$ states are to be considered single-quasiparticle states. The theoretical magnetic moment of such a state in an odd-mass nucleus is that of the corresponding particle state[11] and the Schmidt value is as good an expectation for one as for the other. Our measured moment $\mu(\frac{1}{2}^-) = +(0.58 \pm 0.03)\mu_N$ is indeed not far from the value $\mu(\text{Schmidt}) = +0.64\ \mu_N$.

In closing, I would like to show a photograph (Fig. 12) of a new quartz drive. The design follows from a mutual education process by Dr. Potzel and myself on the subject of the piezoelectric properties of quartz.

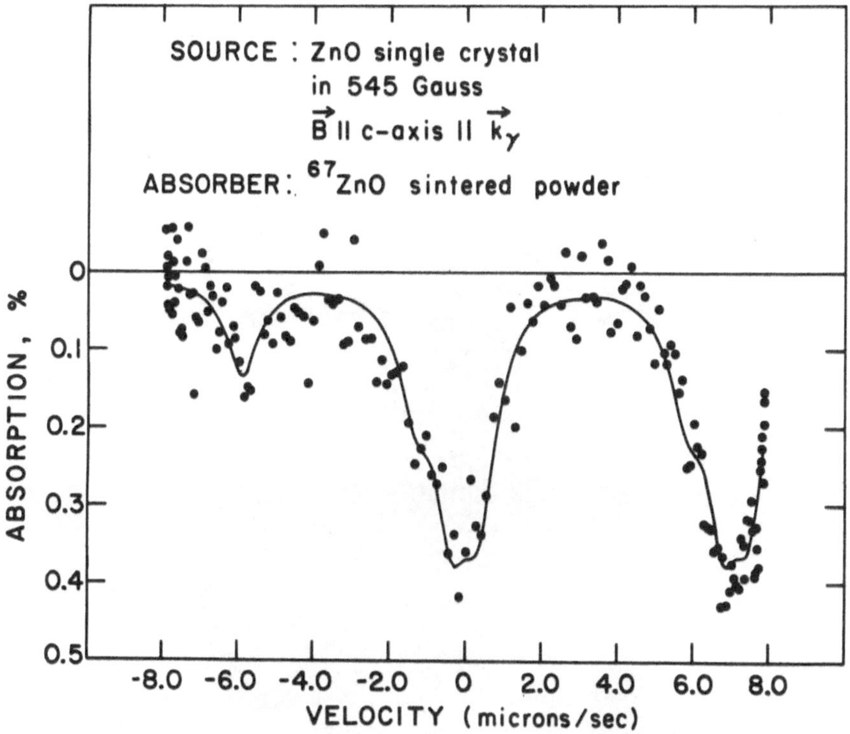

Fig. 11. Spectrum in 545 Oe.

The old spectrometer suffered from lost motion in the joints, from motion in other than the x direction when voltage was applied in the x direction, and from a miscellany of mechanical resonances, some of whose consequences led to periods of alternate excitement and amusement. To be sure, the latter was mainly that of my friends who took part in the excitement and from whom I could not afterwards conceal the prosaic explanations.

The new drive uses the property of expansion in the y direction when voltage is applied along x. It contains many subtleties which will become public knowledge if the apparatus fulfills our expectations.

Fig. 12. New quartz-crystal spectrometer.

† Work performed under the auspices of the U.S. Atomic Energy Commission.

1. P. P. Craig, D. E. Nagle, and D.R.F. Cochran, Phys. Rev. Letters 4, 561 (1960).

2. S. I. Aksenov, V. P. Alfimenkov, V. I. Lushchikow, Yu. M. Ostanevich, F. L. Shapiro, and S. K. Yen, Zh. Eksperim. i Teor. Fiz. 40, 88 (1961) [Soviet Phys. — JETP 13, 63 (1961)].

3. A. I. Beskrowny, N. A. Lebedev, and Yu. M. Ostanevich, Preprint p. 14-5958, Joint Institute for Nuclear Research, Dubna (1971) (unpublished)

4. H. de Waard and G. J. Perlow, Phys. Rev. Letters 24, 566 (1970).

5. H. J. Lipkin, Ann. Phys. (N.Y.) 23, 28 (1963).

6. A. Lurio, Phys. Rev. 126, 1768 (1962).

7. S. C. Abrahams and J. L. Bernstein, Acta Cryst. B 25, 1233 (1969).

8. R. A. Robie and J. L. Edwards, J. Appl. Phys. 37, 2659 (1966).

9. D. von Ehrenstein and J. P. Schiffer, Phys. Rev. 164, 1374 (1967).

10. See, for example, D. J. Rowe, Nuclear Collective Motion (Methuen, London, 1970), p. 200.

11. L. S. Kisslinger and R. A. Sorensen, Rev. Mod. Phys. 35, 853 (1963).

MÖSSBAUER EFFECT - "QUO VADIS?"

Hans Frauenfelder

University of Illinois

Urbana, Illinois, USA

Through the selection of the title, Moshe Pasternak has fixed the direction of my talk. It will consist of two parts,

1. Some remarks about the present and about future conferences
2. The partial construction of a non-Hermitian 3 x 3 matrix.

1. Remarks to the Conference

I was somewhat worried when, after arriving four hours late at Lod airport, nobody knew or expected me either there or at the hotel. And when I found myself in the middle of four Finnish-speaking people, I was no longer certain of being in the right country. However, everything turned out well after that. In fact, I feel now, at the end, that the conference has exceeded expectations. The choice of location was most fortunate. The kibbutz Ayelet Hashahar has been a very pleasant conference site. The food, all ingredients homegrown, has been excellent. The always fresh roses at the speaker's table have added an important component to all talks. There was no possibility of escape from the kibbutz. Consequently, all participants have spent a great deal of time on mutual education, without feeling overworked or bored.

A number of remarks can be made concerning the talks that we heard at this conference. The content of the invited and contributed papers was generally of high quality. However, I would offer the following advice to those organising future

Mössbauer conferences. No speaker should be allowed to carry more than two slides with Mossbauer spectra, and no slide should contain more than 25 words or numbers. Another problem occured here as in any other conference: Speakers know their allotted time at least a few months ahead. Still, when the bell rings, many make a most astonished face and ask for more time. Two classical methods to terminate talks have been practiced in the past. In one, the chairman walks up-and-down, about 50 cm in front of the speaker, holding the alarm clock. In the other, the chairman bodily removes the speaker from the rostrum. I suggest a third approach for the next conference. Speakers will be allowed to continue for as long as they want, provided they pay, then and there, 1 dollar for the first minute, 2 dollars for the second minute, and so on. The money would be invested in champagne for the banquet.

More seriously, in a field that ranges as widely as the Mossbauer effect, future conferences can probably be made even more useful and attractive by considering the following suggestions:-

Invited Papers should be given somewhat more time, say, 50 min. instead of 35 min., but they should predominantly have a review character. Details in one small sector of the entire field are not of sufficient interest to an audience that includes, in addition to nuclear and solid state physicists, also chemists, metallurgists and geologists. On any given subject, the audience as a whole shares little common knowledge. The introductory part of every invited paper should hence always include a presentation of the fundamental facts in a way that is understandable to all listeners. The contributed papers raise another problem. It was physically impossible to present all of them. Selecting about 10% was a painful process both for those performing the selection and those not being selected. The use of rapporteurs, as in high-energy conferences, would probably be a better solution. I was favorably impressed by the session on phase-transitions in which Shenoy acted as a kind of rapporteur, but gave each of the authors of the papers under review a chance to add a few clarifying remarks. It is possible that such an arrangement would make future meetings more exciting and more informative.

The final general remark refers to the title of the

conference ("Applications"). It is not easy to say what applications are and some of the papers certainly do not qualify under this title. However, a number of observations may perhaps bear on those papers that describe truly applications to fields other than physics. A Mössbauer physicist, in his search for new territories to be conquered, often encounters unexpected difficulties (and opportunities) and he has to learn to interact with the scientists already active in this territory. Experience has led to the following rules:

(a) The natives are not as stupid as they appear at first.
(b) Work without a native guide is very often doomed to be wrong, or uninteresting, or both.
(c) A common language (for instance broken English) can be a handicap. The same words may mean very different things in physics and in the applied field. It takes time to establish unambiguous communication with the natives.
(d) The Mössbauer effect is not always the best tool to reach a given goal.
(e) We can often learn from the applied field something fundamental about physics. This point has been employed by Danon.

2. A 3 x 3 Matrix

The non-Hermitian 3 x 3 matrix mentioned earlier is as follows:

	Ideal Mossbauer effect (i)	Mossbauer effect in the real world (r)	Applications (a)
Past (p)	$\langle p \mid M \mid i \rangle$	$\langle p \mid M \mid r \rangle$	$\langle p \mid M \mid a \rangle$
Now (n)	$\langle n \mid M \mid i \rangle$	$\langle n \mid M \mid r \rangle$	$\langle n \mid M \mid a \rangle$
Future (f)	$\langle f \mid M \mid i \rangle$	$\langle f \mid M \mid r \rangle$	$\langle f \mid M \mid a \rangle$

In the folowing I will try to discuss these matrix elements going down each column from the past to the future.

2.1 $\langle p \mid M \mid i \rangle$ and $\langle n \mid M \mid i \rangle$

The two matrix elements $\langle p \mid M \mid i \rangle$ and $\langle n \mid M \mid i \rangle$ are, apart from small corrections, identical: Mössbauer discovered the effect and did not rest till he had explained it fully. Very little has been added since 1958 to what can be called theory and experiment of the ideal Mössbauer effect. I predict with considerable confidence that the future will bring very little here that is new so that $< n | M | i > \approx < f | M | i >$. One remark, however, can be made: The simple theory of the Mössbauer effect is such a good teaching tool that it should be used more frequently to enrich the presentation of quantum mechanics.

2.2 $< p | M | r >$ and $< n | M | r >$

The development of the understanding of the Mossbauer effect in real systems took some time, because nearly all of the physicists who worked early in this field came from nuclear physics. They consequently knew little about real solids. After the first series of exciting experiments, however, most of the holes in the theory were filled in and at present, the basic aspects of the Mossbauer effect in real systems are well understood.

2.3 $< f | M | r >$

The matrix element $< f | M | r >$ is harder to discuss. When Moshe selected the title of my talk, he probably assumed that the virtual existence of prophets in Israel would guide me. I have heard of such guidance in the case of a physics student who was a Rosicrucian. When he solved a very difficult problem, his professor asked him how he had done it. "Oh, it was simple", the student said, "I just asked Newton and Einstein for help". Despite my best efforts, I have not received any such privileged information and I can therefore only guess in which direction future progress will occur:

(a) Better instrumentation. The beautiful work with ^{67}Zn and ^{181}Ta will very likely lead to better spectrometers for all Mössbauer work. The availability of relatively cheap minicomputers will make the data collection and evaluation more flexible.
(b) More Mössbauer nuclides. Step by step, more Mössbauer transitions are added to the list. Most are difficult to use, but some new ones will probably become valuable probes.
(c) High-resolution spectroscopy. Most of us have gone through a strange experience: After unsuccessfully trying to overcome a certain problem for many weeks or months, suddenly a door opens and the problem gets solved, without a new approach. In retrospect it seems difficult to understand why it did not work on the first day. One explanation is that nature likes to be wooed; a more

realistic one is that many small improvements add up to form a crucial step. In any case such a phenomenon has occured in ^{67}Zn and ^{181}Ta, and these nuclides now may become extremely important for the investigation of finer, but conceivably very important, effects. One example has already been given by Kaindl, the temperature dependence of the transition energy. I expect more such effects to turn up in the next few years.

(d) Source and absorber preparation. One crucial ingredient in high-resolution spectroscopy has already been listed under (a), namely, very-high quality spectrometers. The second ingredient is the understanding of what makes good sources and good absorbers. Work with ^{67}Zn and ^{181}Ta will improve this ingredient and the knowledge so gained will, I expect, influence work with other nuclides. In addition, source preparation through implantation may become much more important.

(e) Line-width problems. With the better spectrometers, sources and absorbers, it will be increasingly possible to study not only the "standard" Mössbauer parameters, such as hyperfine splittings and isomer shifts, but also line shapes. Diffusion and relaxation processes thus will become more central. At the moment, we are confronted with one very specific problem, the anomalously small line width observed by Potzel and Perlow in ^{197}Au. Two possibilities exist: At a high-range conference some years ago, a speaker showed a slide with

Two archaeologists and the caption:

IF THIS IS WHAT I THINK IT IS, IT COULD BE
THE DISCOVERY OF THE CENTURY

A year later, another speaker showed the same slide, but the caption had changed:

IF THIS IS WHAT I THINK IT IS, LETS COVER
IT AGAIN AS QUICKLY AS POSSIBLE

More work is needed before it can be decided which caption is more appropriate for the Potzel-Perlow result. In particular, as stressed by Wertheim, the lifetime must be measured with a source that produces a narrow line width.

(f) After-effects. Some of the confusion surrounding after-effects is still with us. I find it particularly hard here to distinguish garbage from gold. I believe that more experimental work with simple systems is needed. The inert-gas matrix technique may be particularly useful. A critical review on after-effects is needed. It should summarize what is known and what is uncertain and combine the information from Mössbauer effect, perturbed angular correlation, and recoil methods.

(g) Coherent processes. Coherent processes will probably be studied more deeply in the future. The theory of Kagan and Afanasjev[1], and the corresponding experiments from a basis on which future work can be built. Rayleigh scattering of resonant gamma rays, as discussed by O'Connor, very likely will also become a more widespread tool.

(h) Combinations. The combination of Mossbauer measurements with other tools, such as nuclear orientation and perturbed angular correlation, may help to elucidate some problems.

2.5 $\langle n | M | a \rangle$

The list of applications has grown enormously in the past ten years or so, as a glance at any recent conference proceedings[2] and at the list of contributed papers here will show. The Mössbauer effect has been introduced to fields as different as limnology and the motion of ants. Moon rocks and Greek vases have been examined. A detailed evaluation of where we stand today is not really feasible in one talk, but various papers at this meeting have provided at least a partial review.

2.6 $\langle f \mid M \mid a \rangle$

Back again to the role of prophet: What will happen in the future? Here it is even harder to guess than in the case of the element $\langle f \mid M \mid r \rangle$, but I expect the following developments:

(a) Investigations in fields such as metal physics and biological molecules will become more careful, more systematic, and the interaction with other techniques will become stronger. The "quick buck" artists will drop out and leave the work to the dedicated "long distance runners".
(b) The Mössbauer effect will increasingly be applied to systems such as ferromagnetic superconductors, liquid crystals, glasses, sols and thin films. Remark (a), of course, applies here also. Phase transitions in such systems will be of particular interest.
(c) "High-resolution" nuclides, (^{67}Zn and ^{181}Ta), will be applied and some unexpected results will probably show up. While nearly all the applied experiments up to the present time use the small line widths, Γ, of recoil-free transitions, it seems likely that a few of the future experiments will make use of the high Q values ($Q = E_\gamma/\Gamma$).
(d) Radiation-induced and radiation-related effects will probably be studied in more detail in various systems.
(e) Surprisingly, the Mössbauer effect has found little use in technology, for instance in measurements of quality control and so on. It is likely that the general trend towards "relevant" science will induce some Mössbauer physicists to find more practical applications.

3. Summary

The field of the Mössbauer effect is not dead, but surprisingly active. Much of the original fever-pitch excitement is gone, but it has been replaced by serious, careful, and dedicated work. The field shares one problem with other mature and large areas of physics. It is hard to recognize the few truly outstanding and novel contributions in the flood of papers. The conference that we have just attended has helped in this chore and I hope that future meetings will be called, with the same goal, to spotlight the progress and separate the essential from the rest.

REFERENCES

1. A.M. Afanasjev and Yu. Kagan, Soviet Physics, JETP 25, 124 (1967) and other references quoted therein; A.M. Afanasjev and Yu. Kagan, Soviet Physics, JETP 21, 215 (1965).

2. Proceedings of the Conference on the Applications of the Mössbauer Effect, (1967), Editor, I. Deszi - Akademiai Kiado, Budapest 1971.

LIST OF CONTRIBUTED PAPERS

Demonstration of Parity Violation in the Beta Decay of ^{61}Co by Mossbauer Effect, K.P. Schmidt, R. Coussement G. Langouche, J.P. Lafaut.

The Hyperfine Asymmetry Effect, R. Coussement, M. Rots, G. Langouche, K.P. Schmidt.

Mossbauer Effect Evidence for the Coexistence of Ferromagnetism and Superconductivity, R.D. Taylor.

Time Dependent Hyperfine Interactions in α-Fe_2O_3*, B. Balko and G.R. Hoy.

MB - Experiments with Conversionelectrons on ^{182}W and ^{145}Pm, H. Bokemeyer, K. Wohlfahrt, E. Kankeleit.

Measurement of 2s and 3s Electron Spin Density in Iron Metal and Fe_2O_3, C.J. Song, J.M. Trooster, N. Benczer-Koller.

Observation of Narrow Absorption Lines in Mossbauer Experiments on ^{197}Au, W. Potzel and G.J. Perlow.

Quadrupole Moment of the First Excited State in ^{133}Cs from the Mossbauer Effect in Cesiated Pyrolytic Graphite, L.E. Campbell, G.L. Montet, and G.J. Perlow.

On the Perturbed Angular Correlation of Scattered Mossbauer Gamma Rays, R. Ehehalt, E. Nolte, H.J. Korner, F.E. Wagner, P. Kienle.

Mossbauer Experiments at Very Low Temperatures, M.T. Hirvonen and T.E. Katila.

Mossbauer Studies of Apollo Lunar Samples, T.C. Gibb, R. Greatex, and N.N. Greenwood.

Dissolution of Metastable Precipitates in Cu-Fe Alloys, by Plastic Deformation, F. Hornstein, M. Ron.

Study of Hexagonal Strontium Aluminium Ferrites by Means of the Mossbauer Effect, V. Florescu, D. Barb, M. Morariu, D. Tarina.

Contributions to the Mossbauer Line Shape in Coupled Hyperfine Interactions, D. Barb, S. Constantinescu, L. Diamandescu and D. Tarina..

Mossbauer Study of Diffusion of Exchangeable Cobalt(II) - Ions on the Surface of Vermiculite, J. Helsen, K.P. Schmidt, G. Langouche, R. Coussement, T. Meykens, and T. Cahkupurakal.

Electric Quadrupole Interaction (EQI) Studies in Hf Based Ceramics, P. Boolchand, Ching-lu Lin, S. Jha and F.L. Koucky.

Mossbauer Study of Iron-bearing Tourmalines, E. Hermon, D. Simkin, G. Donnay and W.B. Muir.

Mossbauer Analysis of Lake Sediments, J.M.D. Coey.

Mossbauer Spectrscopyin the Study of Fossils, E. Mattievich, J. Danon

Mossbauer Scattering: Promise and Limitations as a Quantitative Metallographic Tool for Steels, L.H. Bennett and L.J. Swartzendruber.

Mossbauer Effect in Catalase, Y. Maeda, A. Trautwein, U.Gonser

Mossbauer Studies on the Electronic Structure of Ferrous Iron in Hemoglobin, A. Mayer, G. Wagner, H. Ubelhack, H. Formanek, F. Parak, -. Schlecht, and H. Eicher.

Experimental Solution of the Phase Problem: Freezing of Protein Single Crystals Without Destruction and Development of a Multicounter System, F. Parak, U. Biebl, U.F. Thomanek, and R.L. Mossbauer.

Effects of Ligand Nuclear Magnetic Moments in Mossbauer Spectra of Iron Proteins, K. Spartalian and W.T. Oosterhuis.

The Study of Intermolecular Interactions in Phthalocyanines using Mossbauer Spectroscopy, J.L. Przybylinski, A. Nath.

Mossbauer Spectra of Bacterial Catalase and Catalase Compound II, L. May, Merle H. Arnold, and C.T. Kuo.

Mossbauer Spectroscopy of Hepatic Microsomes, L. May and J.L. Holtzman.

Can Mossbauer Spectroscopy Detect Fe(IV) in Hemoproteins? L. May.

Mossbauer Spectroscopy of S=5/2 Iron at Sites of Low Symmetry in Boimolecules: Application to Conalbumin, G. Lang, R.C. Woodworth, P. Aisen.

Pressure Dependence of the Isomer Shifts and Quadrupole Splittings of Di^{-} and Trivalent Europium Compounds, G. Wortmann, U.F. Klein, W.B. Holzapfel, and G.M. Kalvius.

Chemical Shifts of Core-Electron Binding Energies in Os Compounds, G. Kaindl.

Systematics of Isomer Shifts for Impurities of ^{197}Au and ^{99}Ru in Transition Metals, G. Kaindl and D. Salomon.

Systematics of Isomer Shifts in Dilute Alloys and in Compounds of Transition Metals, G. Wortmann, F.E. Wagner, and G.M. Kalvius.

Pressure dependence of Gamma Ray Energy Shift for Fe^{57} in Cu lattice at 298 and 94°K, S.S. Nandwani, and S.P. Puri.

Saturation of Mossbauer Absorption Spectrum Areas in Thick Absorbers, J.M. Williams and J.S. Brooks.

Mossbauer Studies on the Role of Time Factor During the X-Rays Radiolysis of Solid Ammonium Trioxalatoferrate Trihydrate, C. Kellersohn, F. Soubiroux and C. Hubert.

Comparative Mossbauer Study of the Chemical After Effects of Various Nuclear Processes in Tin (II) Sulphate, J.M. Friedt, W. Vogl.

Mossbauer Study of the Ordered Fe_3Ga Alloy, Y. Nakamura.

Studies of Relaxation Phenomena in Paramagnetic Dy Compounds at High Temperatures, A. Almog, E.R. Bauminger and S.Ofer.

The Orientation-Dependence of Atomic Spin Relaxation Deduced from Mossbauer Spectra of Magnetized Fe $NH_4(SO_4)_2 \cdot 12H_2O$, B. Braunecker and H.H.F. Wegener.

Investigation of Hyperfine Interaction and Structure in $FeSiF_6 \cdot 6H_2O$ By Mossbauer Measurements, H. Spiering, R. Zimmermann, G. Ritter.

Mossbauer Spectra of the Eight-Coordinated Complex Tetrakis-(1,8-Naphthyridine)Iron(II) Perchlorate, R. Zimmermann, H. Spiering and G. Ritter.

An Investigation by Mossbauer Spectroscopy of Fe^{57} in Pure and Oxidized Tin Chalcogenides, V. Fano, I. Ortalli.

Sn-Pd Characterization by Mossbauer Spectroscopy, R.L. Cohen and K.W. West.

On The Isomer Shifts in Dy-Transition Metal Intermetallic Compounds, A.A. Gomes, A.P. Guimaraes, J. Danon.

Mossbauer Effect Evidence for a Covalent Te-Se Bond, P. Boolchand, P. Suranyi.

Inter and Intramolecular Bonding Effects in Organotin Compounds, R.H. Herber.

Mossbauer Studies of Pentlandite and Related π-Phases, O. Knop and F.W.D. Woodhams.

Structural Investigations in Hafnium Fluorides, P. Boolchand, Ching-Lu Lin, S. Jha, and Frank L. Koucky.

The Quadrupole Coupling in α-Fe, T.E. Cranshaw.

Properties of Iron Impurities in Beryllium from Mossbauer Studies, C. Janot, P. Delcroix and M. Piecuch.

Mossbauer Effect in $FeCr_2S_4$, A.M. VanDiepen.

High-Resolution Mossbauer Spectroscopy of ^{161}Dy Applied to the Three-Dimensional Ising Crystal, $DyPO_4$, D.W. Forester, W.A. Ferrando.

Hyperfine Interaction of ^{181}Ta in Nickel, G. Kaindl and D. Salomon.

Evidence for the Existence of Non-Equivalent Impurity Sites from Mossbauer Spectra of Implanted $I^{125}\underline{Fe}$ Sources, N.S. Wolmarans, H. deWaard and S.R. Reintsema.

Hyperfine Interactions with ^{57}Fe in Iron Alloys, T.E. Cranshaw.

A Mossbauer Study of Fe_3O_4 - γ Fe_2O_3 Solid Solution, H. Annersten.

Non-Axial Hyperfine Interactions for ^{237}Np in Hexavalent Neptuynl Salts, B.D. Dunlap, I. Nowik, and D. Cohen.

Diamagnetically Substituted Iron Oxides - The Central Peaks, J.M.D. Coey.

Internal Magnetic Fields in $Hf_xZr_{1-x}Fe_2$ Alloys, S. Jha, Ching Lu Lin, M. Blizzard and P. Boolchand.

Mossbauer Study of Ordering in Fe-Si Alloys, L. Haggstrom, L. Granas, R. Wappling and S. Devanarayanan.

Mossbauer STudy of Magnetic Ordering in Cu_2FeSnS_4, U. Ganiel, E. Hermon and S. Shtrikman.

Hyperfine Field at Sn Sites in Sn-doped Cu_2MnIn, W. Leiper and C.C.M. Campbell.

The Disturbance Produced in an Iron Lattice by Cr Atoms, T.E. Cranshaw.

A Study of Hyperfine Fields in Heusler Alloys, J.S. Brooks, and J.M. Williams.

Determination of Magnetic Properties of Uranium Monotelluride from Mossbauer Effect and Neutron Diffraction Measurements, G. Longworth, M. Kuznietz, and F.A. Wedgwood.

Mossbauer Studies of Electron Irradiated Ferricyanides, E. Baggio Saitovitch, D. Raj and J. Danon.

Evidence of ^{57}Co Aggregation in the Grain Boundaries, G. Martin.

Radiation Damage in Iron at Low Temperature Studied with the Mossbauer Effect, W. Mansel, J. Prechtel, A. Schaefer, G. Vogl, W. Vogl.

Mossbauer Studies of Thermal Decomposition of Alkali-Ferricyanides, D. Raj, E. Baggio Saitovitch and J. Danon.

Chemical Consequences of Various Nuclear Transformations Feeding Mossbauer Transitions in ^{151}Eu and ^{153}Eu, J.M. Friedt, G.M. Kalvius, F.E. Wagner, U. Zahn.

Chemical Effects of the $^{57}Co(EC)^{57}Fe$-Reaction in Solid Cobaltihexacyanides, J. Fenger and K.E. Siekierska.

Irradiation Induced Disorder in a CoPt Alloy, A. Simopoulos, G. Vogl, W. Vogl.

Anomalous Spin States of Iron in Mossbauer Emission Spectra of ^{57}Co-Labelled Cobalt Complexes, J. Ensling, B.W. Fitzsimmons, J. Fleisch, P. Gutlich, K.M. Hasselbach.

Chemical Nature of Electron Capture (EC) Decay Products in Chloride and Cyanide Complexes of ^{57}Co, J.M. Friedt, G.K. Shenoy, G. Abstreiter.

Mossbauer Studies of Cu-Fe Thin Films, D.L. Williamson, J. Lauer, and W. Keune.

Octahedral and Tetrahedral Interstitial Carbon Atoms in Iron and Steel, F.E. Fujita.

Mossbauer Measurements on the Bipyramidal Lattice Site in $BaFe_{12}O_{19}$, E. Kreber and U. Gonser.

Low Temperature Mossbauer Measurements on Ferroelectric Boracites, J.M. Trooster.

The Magnetic Structure of Invar, B. Window.

Mossbauer Study of Yb^{3+} Ions in $YbFeO_3$ and Yb-Ethylsulfate, F. Gonzalez Jimenez, P. Imbert and J.D. Jujica.

Mossbauer Study of ^{57}Fe in Doped CaO and $KmgF_3$, J.R. Regnard.

Mossbauer Relaxation Spectra of Yb^{3+} in a Cubic Site, F. Gonzalez Jimenez, P. Imbert.

Paramagnetic Hyperfine Interactions in a d^2 Complex of Fe^{57}, W.T. Oosterhuis, F. de S. Barros.

Paramagnetic Hyperfine Interactions in d^4 Complexes of ^{57}Fe, E.A. Paez and D.L. Weaver.

Mossbauer Study of Diffusion in Liquid Glycerol, J.G. Mullen.

Goldanskii-Karyagin Effect vs. Preferred Orientations (Texture) H.D. Pfannes, U. Gonser.

Crystallographic Phase Transition in $FeCl_2$,2py, J.P. Sanchez L. Asch, J.M. Friedt.

Mossbauer Recoil-Free Fraction in $^{119}Sn(C_6H_5)_4$ - Doped $Ge(C_6H_5)_4$, $Sn(C_6H_5)_4$, and $Pb(C_6H_5)_4$ Matrices, H. Sano and Y. Muto.

Phase Transitions in $(NH_4)_3FeF_6$, S. Morup and N. Thrane.

Internal Motion Effects in the Mossbauer Spectra of Ferrous Fluosilicate, M.H. Villavicencia, P.H. Domingues and J. Danon.

Anisotropy of the Debye-Waller Factor for the Cesium Ion of C_8Cs in Pyrolytic Graphite, L.E. Campbell, G.L. Montet, G.J. Perlow.

Mossbauer Spectrum of 2d and 3d Anisotropical Crystals, Y. Imry, B. Gavish.

SUBJECT INDEX